Learn & Leverage

TECHNOLOGY

For Real Estate Investors & Business Owners

the savvy LANDLORD

By Steven VanCauwenbergh

Learn & Leverage
TECHNOLOGY
For Real Estate Investors & Business Owners

International Standard Book Number: 978-1-943186-04-4

First printing 2015

Printed in the United States of America

Library of Congress Cataloging-in-Publication Data is available from the publisher

Published by Teflon Publishing, 8 NE 48th Street, Oklahoma City, OK 73105

Savvy Tech DVD video has 69 videos and is over 2.5 hours in length. There are 4 videos in each chapter as detailed below.

This project has taken over 80 hours of labor to complete! Enjoy my fellow Savvy Investor!

Chapter 1 1. **BackBlaze** 2. **BatchGeo** 3. **Hubzu** 4. **Postlets**	Chapter 10 1. **CrimCheck** 2. **JoinMe** 3. **OhDontForget** 4. **CreditKarma**
Chapter 2 1. **Vumber** 2. **Shoeboxed** 3. **Cert Mail** 4. **Zaba**	Chapter 11 1. **ChromeStore** 2. **Google-Voice** 3. **pdf2jpg** 4. **WeTransfer**
Chapter 3 1. **LandlordProtection** 2. **MyCheapCoupons** 3. **RingCentral** 4. **CrimeReports**	Chapter 12 1. **HundredZeros** 2. **Pixlr** 3. **Lynda.com** 4. **Deal Catcher**
Chapter 4 1. **Buildium** 2. **Appfolio** 3. **Property Manager Cloud** 4. **Eppraisal**	Chapter 13 1. **Grasshopper** 2. **RelayRides** 3. **Bill.com** 4. **Google Calender**
Chapter 5 1. **FindCompsNow** 2. **VoiceConnect** 3. **GoalsOnTrack** 4. **Weebly**	Chapter 14 1. **LendingClub** 2. **Prosper** 3. **eVoice** 4. **Elance**

Chapter 6 1. **Fiverr** 2. **MySmartMove** 3. **PatLive** 4. **AirBnb**	Chapter 15 1. **ODesk** 2. **Mashable-Life Hacker** 3. **Hive** 4. **BackPage**
Chapter 7 1. **BookBub** 2. **GoSection8** 3. **AC Wholesalers** 4. **MetroFax**	Chapter 16 1. **SkillShare** 2. **Rev** 3. **Life360** 4. **SlimInvoice**
Chapter 8 1. **DropBox** 2. **PassPack** 3. **withoomph** 4. **Google-Drive**	Chapter 17 1. **TrueCar** 2. **Amazon Prime** 3. **Dymo-Printer** 4. **PaperPort14**
Chapter 9 1. **ClearNow** 2. **Auction** 3. **Evernote** 4. **Homewyse**	Chapter 18 1. **SnapScan**

savvy INVESTORS .com

01 - BackBlaze.com

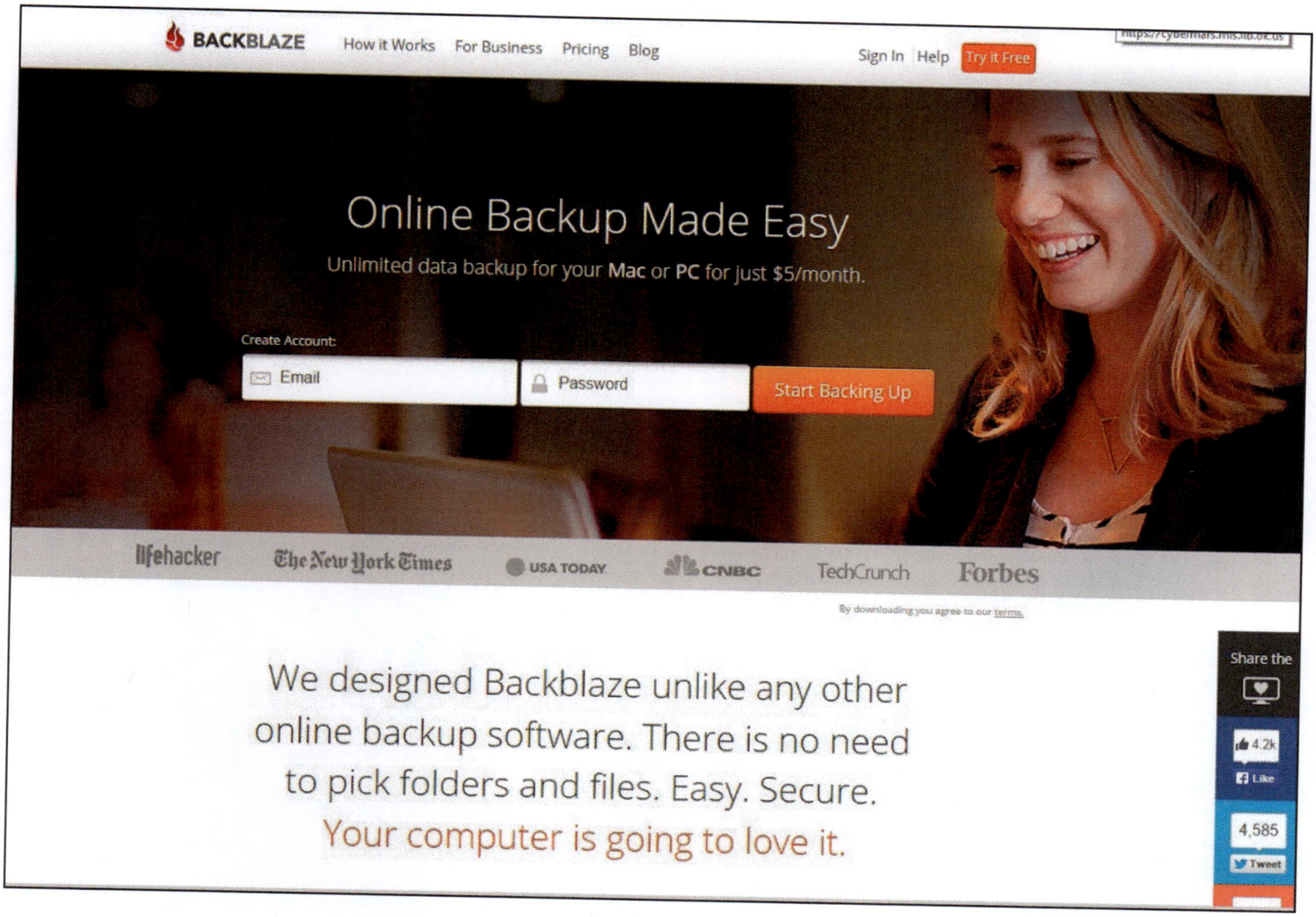

Backblaze is an online backup tool that allows Windows and Mac OS X users to back up their data to an offsite data center. This service is designed for end-users, providing unlimited storage space and supporting unlimited file sizes.

Backblaze is a great way to get all your files backed up automatically. It's easy, It's secure, and if you lose data for any reason, you will be glad you have **Backblaze** backing up all of your files on the cloud.

Award winning unlimited online backup for only $5/month.

savvy INVESTORS .com

02 - BatchGeo.com

BatchGeo.com creates Google Maps from your company's data. It's as simple as copying your company's spreadsheet data into the BatchGeo.com tool. It accepts addresses, intersections, cities, states, and postal codes. BatchGeo creates maps to see where your data lives in the real world. This service can be used to advertise open houses, rental properties, or homes for sale.

Features of BatchGeo.com:

- Easy to make maps with your own data.
- Include lots of information including: property address, listing agent, listed price, square footage, year built, annual taxes, photo of property, and to link to property website.
- Helps potential buyers or tenants find the location of your property.
- Provides a large amount of information to a customer with just one click.
- Helps you track trends for your properties within an area.
- Available App for Apple and Android devices.

03 - Hubzu.com

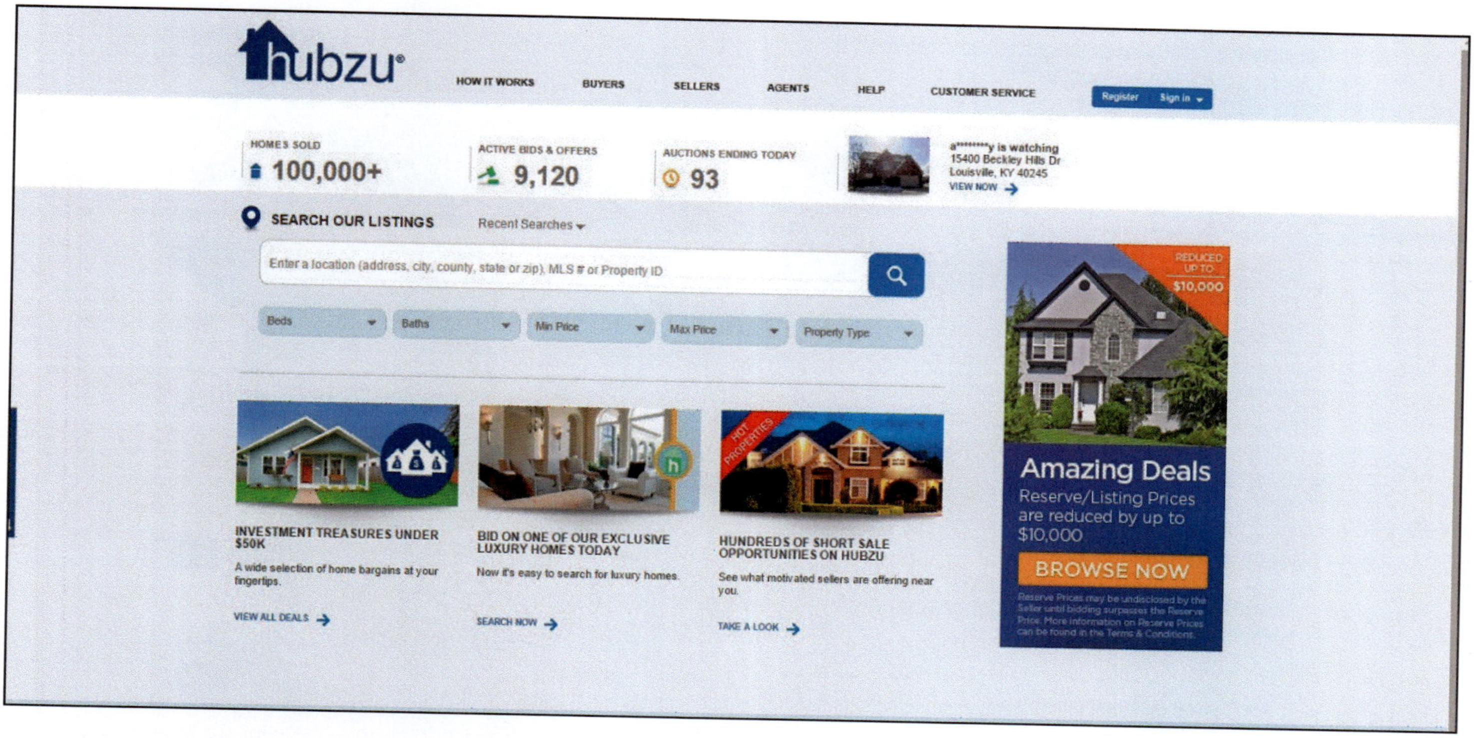

Hubzu.com is an online real estate marketplace you can browse thousands of homes from coast to coast. You can both buy and sell properties through **Hubzu**.com. They sell foreclosed, FSBO, and short sale homes on the site. Search and bid on properties from your computer anywhere in the country at anytime..

Key features of **Hubzu**.com:

- Search and location filters make finding properties in your area simple.
- Register for a free "My **Hubzu**" account to save and watch properties that interest you.
- Get advanced notice when new properties are being added through your "My **Hubzu**" account.
- Schedule a visit of the property through the site or your realtor.
- Two ways to buy: Bid online in real time or make an offer on the property and get a response from the seller.
- An "own it now" price is typically listed. You can purchase the property at anytime at that price.
- Your realtor can represent you and be paid by the **Hubzu** seller.
- When a property is won the buyer is assigned a sellers representative and given their direct contact information.
- Most properties do have a fee associated with buying but it is calculated into your bidding price.

This auction site allows for a more traditional sale where you can work with a realtor and see the property before bidding.

04 - Postlets.com

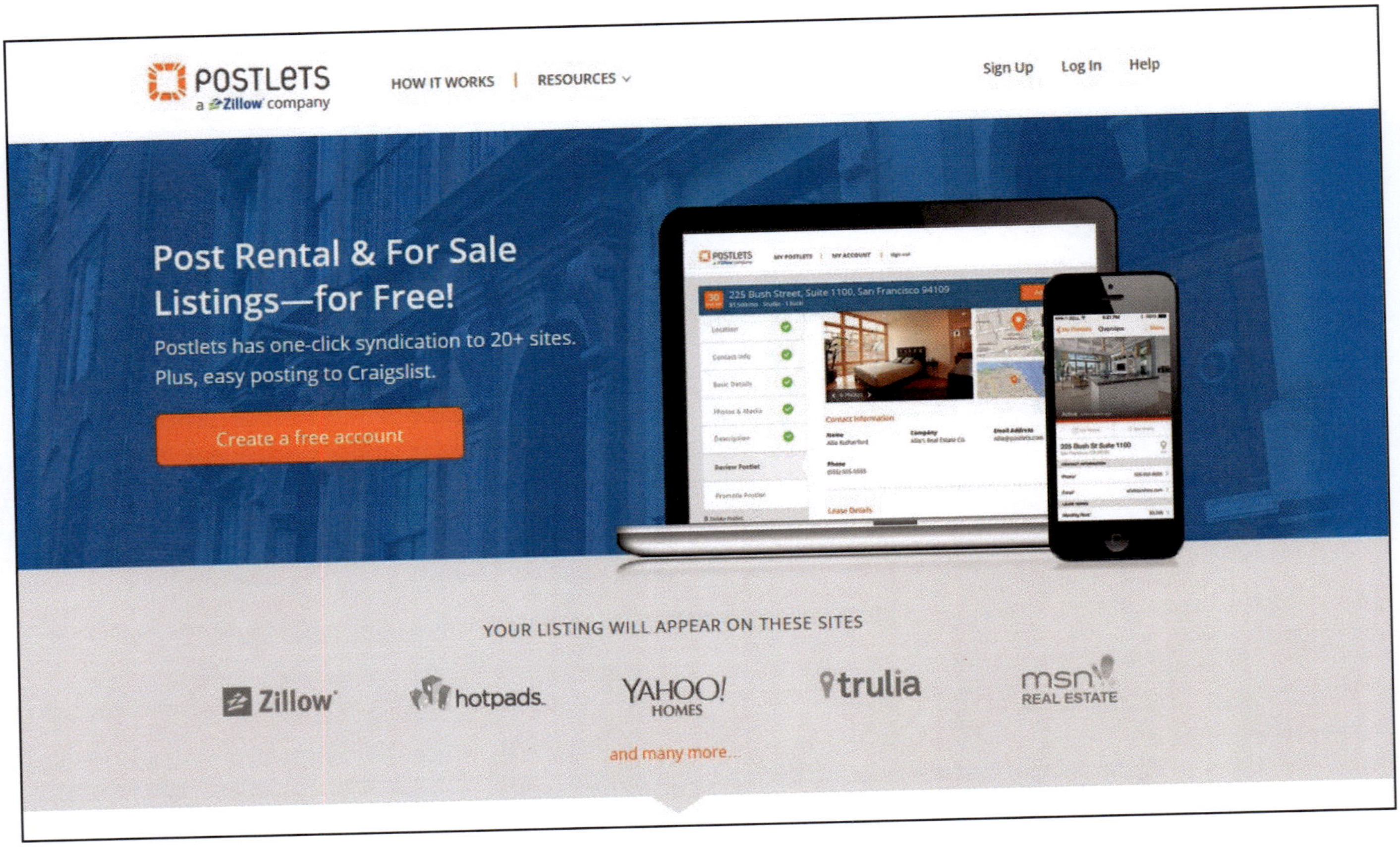

Postlets is an online listing creation and distribution platform. **Postlets** is used by individual real estate agents, property managers and landlords to market their listings to search networks and social media sites.

Postlets is owned by Zillow .

You can post your listing to over 20 different websites with one click of the button.
Keeps your listing in one place, manage and edit.

Founded in 2005, **Postlets** allows individual real estate agents, property managers and landlords to create listings and send them to 13 real estate and social media websites. The free platform itself makes syndication of listings easy!

05 - Vumber.com

Vumber is a virtual phone number – now you can have several numbers on a single phone. With **Vumber**, choose any area code you want and link it to your home, cell, or work phone.

Vumber gives you the ability to add a new phone number to any phone that you own.
You can make and receive calls from your new number without affecting your existing numbers in any way.

Vumber works on ANY phone. Customers can call you on your phone without knowing your real phone number on that device.

You can call tenants back and they will see your new **Vumber** displayed in their Caller ID display.

Vumber allows you to run your business from your existing cell phone or landline, safely and securely. Your phone number will not be revealed to your Customers/Tenants.

Why **Vumber**? This service is great for marketing, bandit signs, and running tests.

06 - Shoeboxed.com

With **Shoeboxed** you can scan, mail, upload or send a picture of a receipt to **Shoeboxed**, and it will store it for you. You can then go online to your account, assign a category, and export a nice looking file when you need it. It's helpful for tracking reimbursements, expense accounts, etc. **Shoeboxed** also has iPhone/iPad/Android apps that are fast and, an easy way to turn receipts into an organized, IRS-accepted archive online.

Shoeboxed is the fast, painless way to digitize and archive receipts and important paper documents in a single, secure location. **Shoeboxed** scans receipts, business cards and other documents and turns these scanned images into extracted, human-verified data that is categorized, organized, fully searchable and available anytime, anywhere. **Shoeboxed** users are able to streamline **accounting** and bookkeeping, maximize every tax deduction, and email expense reports in seconds. **Shoeboxed** also works with all of the tools you already use and know. Works seamlessly with QuickBooks, Xero, Wave Accounting, Outright, ScanSnap, Evernote, and dozens of other programs.

Founded in 2007, **Shoeboxed** now serves over 800,000 users in over 100 countries worldwide

savvy INVESTORS .com

07 - CertifiedMailLabels.com

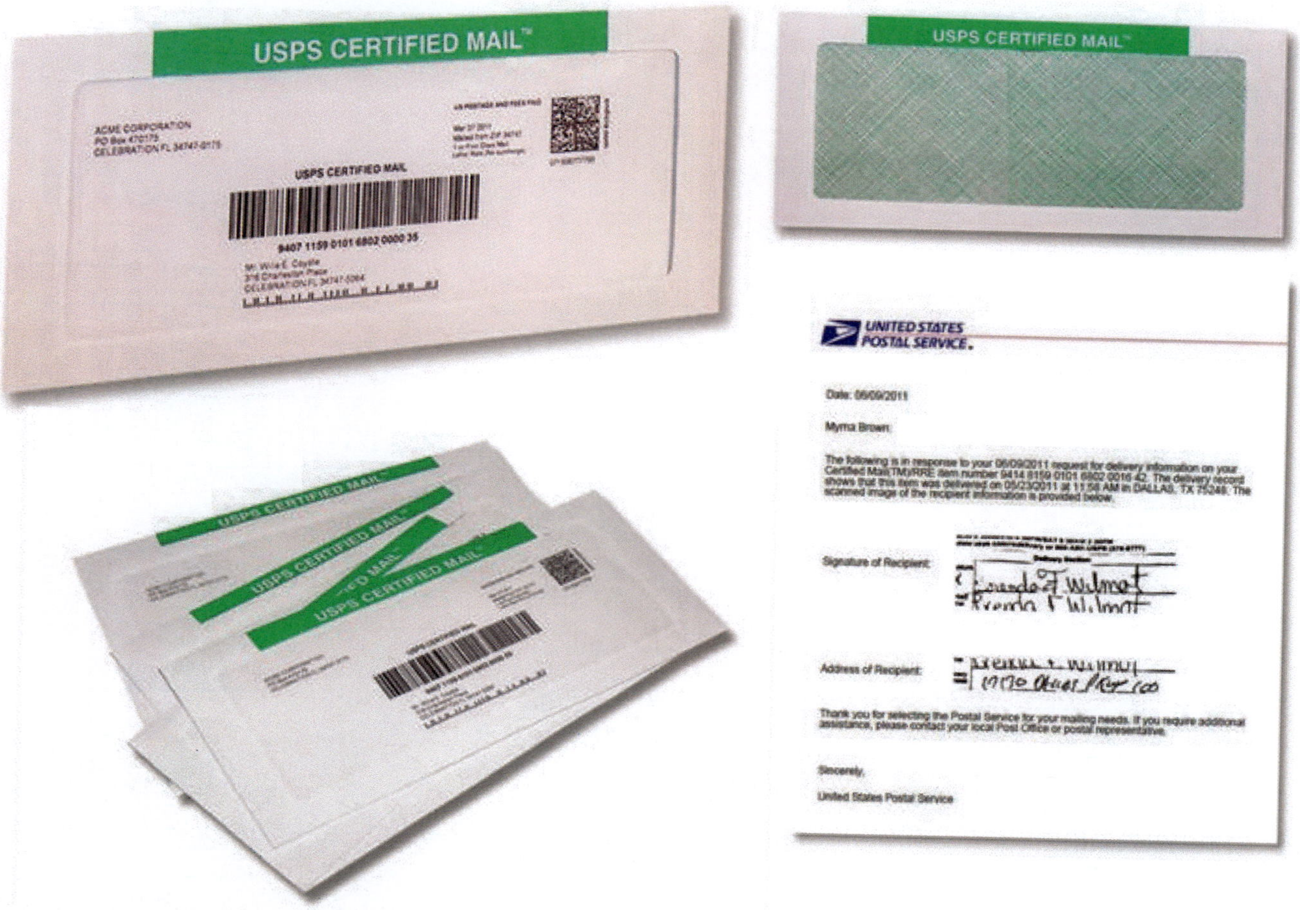

Skip the trip to the Post Office and address and print USPS Certified Mail Labels online! You will save $1.97 postage on each Certified Mail green card receipt. No monthly fees, no long term contracts and no software or special equipment. Get email notification with Electronic Delivery Confirmations, Return Receipt Signatures and 10 year tracking and archive at no extra cost. You'll have proof of mailing, tracking, plus delivery confirmation for all your compliance reporting available 24/7 – 365 days a year.

savvy INVESTORS .com

08 - ZabaSearch.com

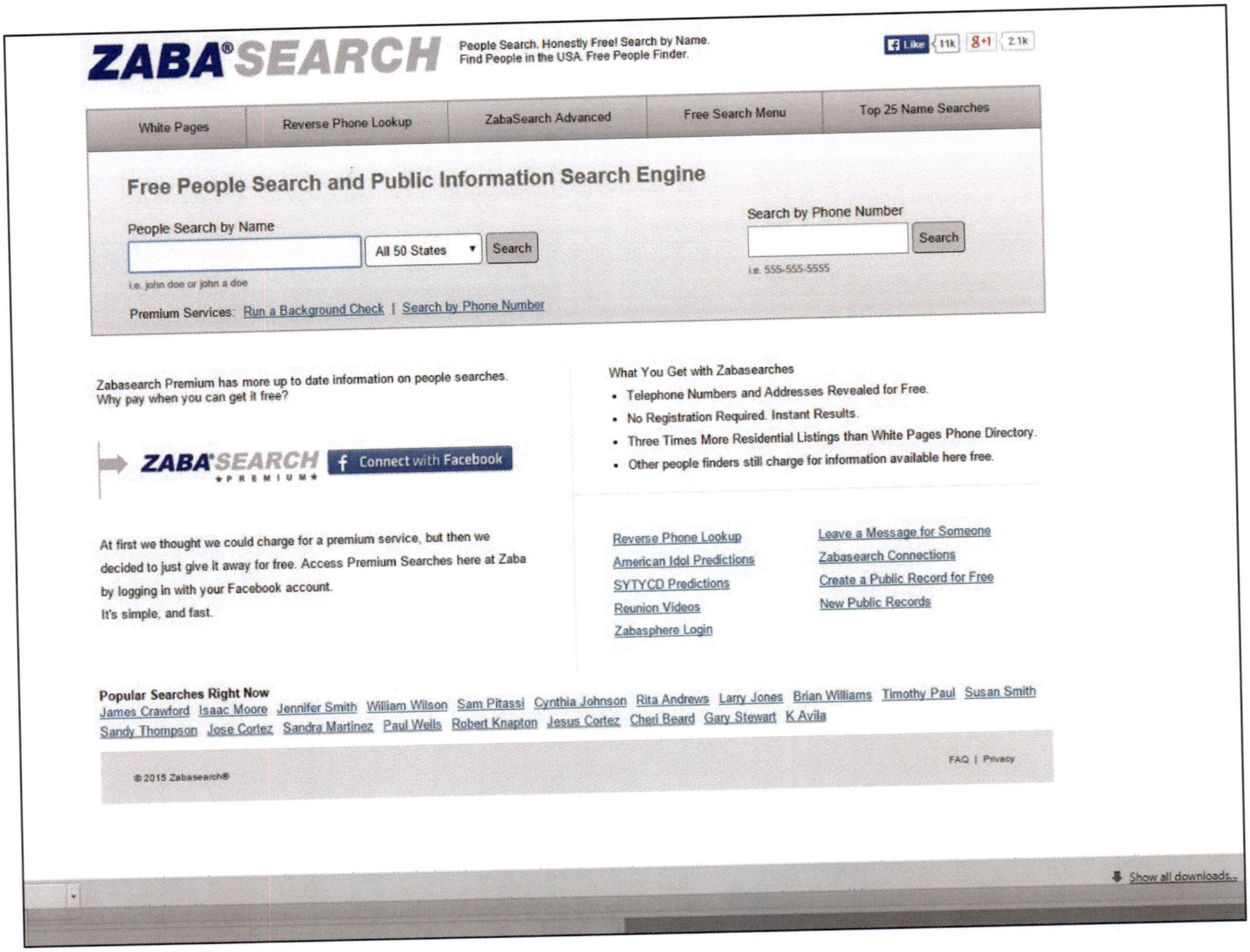

ZabaSearch.com is a people search engine. It allows you to easily search for information about people using their name, phone number, or white pages listings. The data base on ZabaSearch.com has three times more listings than the White Pages Phone Directory. Basic searches are totally free, but you must sign in with your Facebook account to access premium searches - either way there isn't a fee!

Features of ZabaSearch.com:

- Easily find contact information for others.
- Free information that other companies charge for.
- Searches yield telephone numbers and addresses when available.
- Has a reverse phone number lookup tool.

09 - TheLPA.com

The **Landlord Protection Agency®** website went live online in the year 2000. It has helped thousands of landlords overcome multitudes of tenant problems while enabling them to perform more easily and efficiently in management of their rental property and dealing with Tenants.

The goal of The **Landlord Protection Agency®** is to uphold a philosophy and code of ethics among landlords, real estate brokers and rental agents in dealing with Tenants. The LPA caters to landlords in all of the United States, Canada and many other English-speaking countries.

10 - MyCheapCoupons.com

MyCheapCoupons.com! Your source for cheap Lowe's, Starbucks, and now many other coupons!

Get your coupons instantly for immediate printing (if they are available for printing at home), or they will mail them to you in 1-2 business days!

11 - RingCentral.com

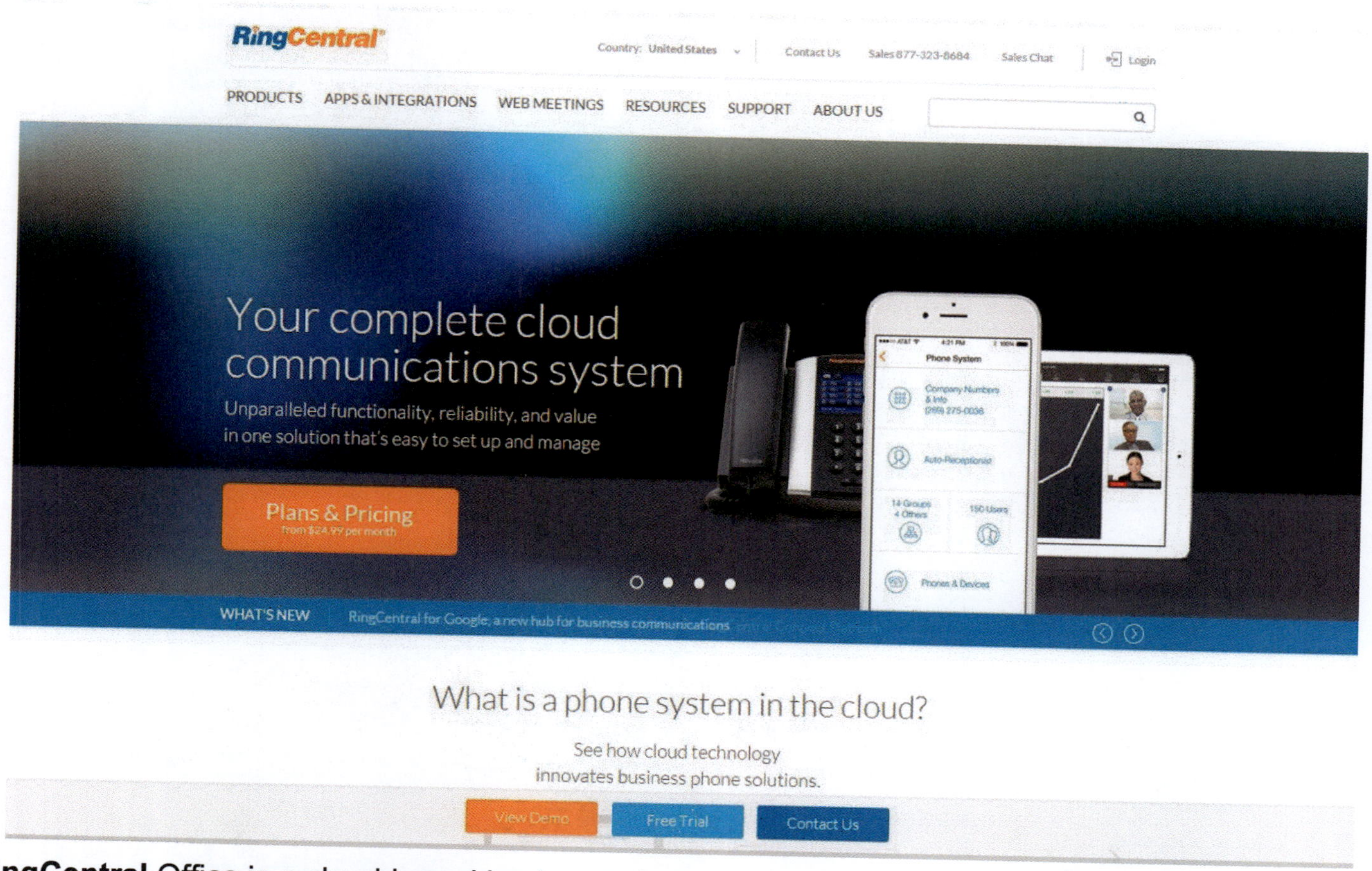

RingCentral Office is a cloud-based business phone and fax system. It combines a hosted phone system with advanced call and fax management, reliable phone and fax service, and full-featured IP phones that arrive pre-configured and ready for use.

With **RingCentral** Office, you can easily connect your office, remote and mobile employees under one phone system, regardless of their location.

Unlike expensive conventional phone systems, **RingCentral** Office is purchased, activated, setup and managed online, which means they manage it and you use it. By eliminating the need for complex hardware, lengthy installation and technical expertise, you get the functionality and flexibility to easily connect with customers without high costs or hardware hassle.

With one flat, monthly fee, no contracts and zero setup costs, **RingCentral** Office makes your business communication easy, affordable and accessible from your home, office, or wherever you need to be.

savvy INVESTORS .com

12 - CrimeReports.com

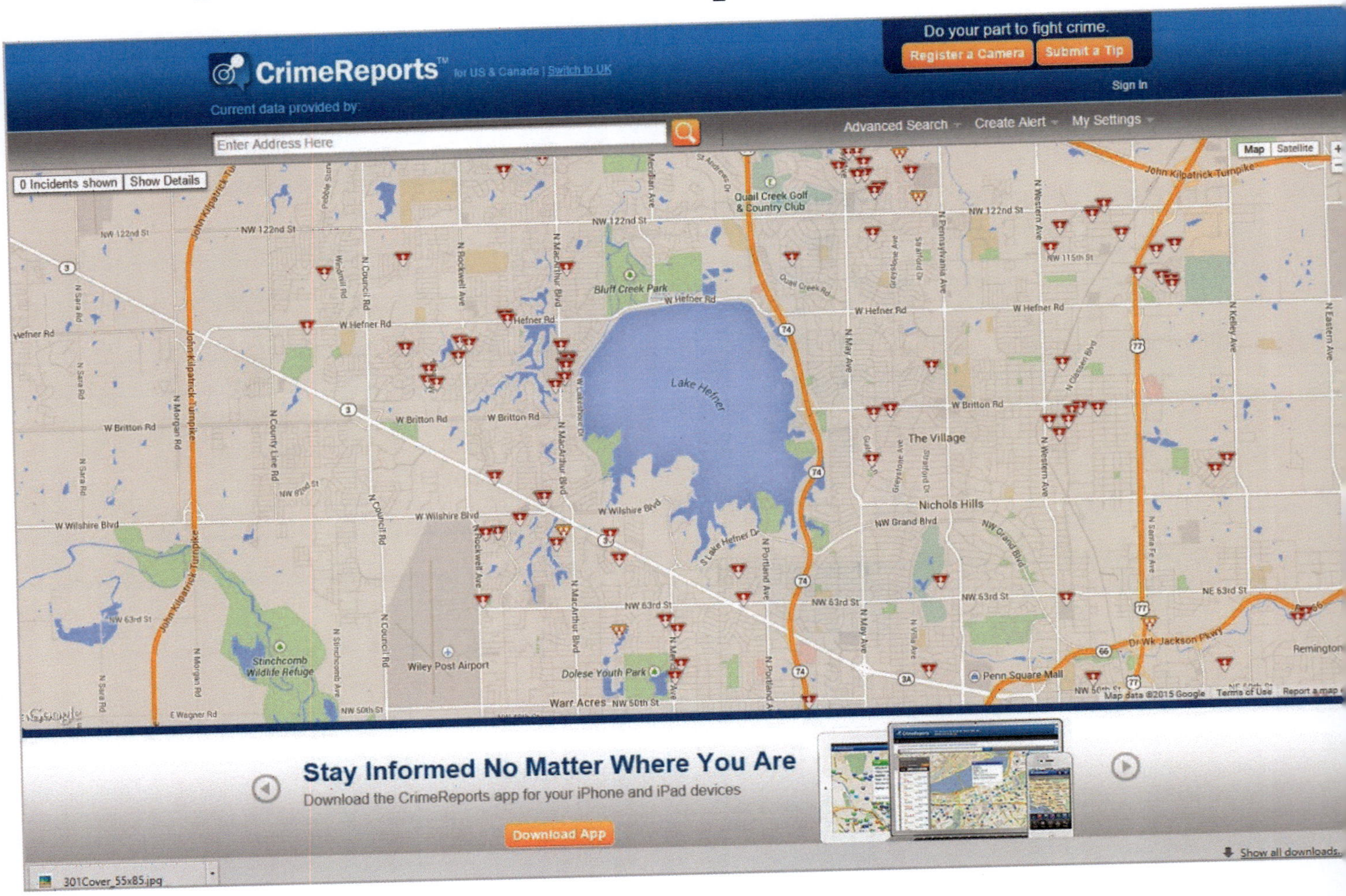

CrimeReports™ auto-publishes crime data.. You can access and browse crime information as shown on a map from any computer or mobile device and are able to submit anonymous tips through the platform.

13 - Buildium.com

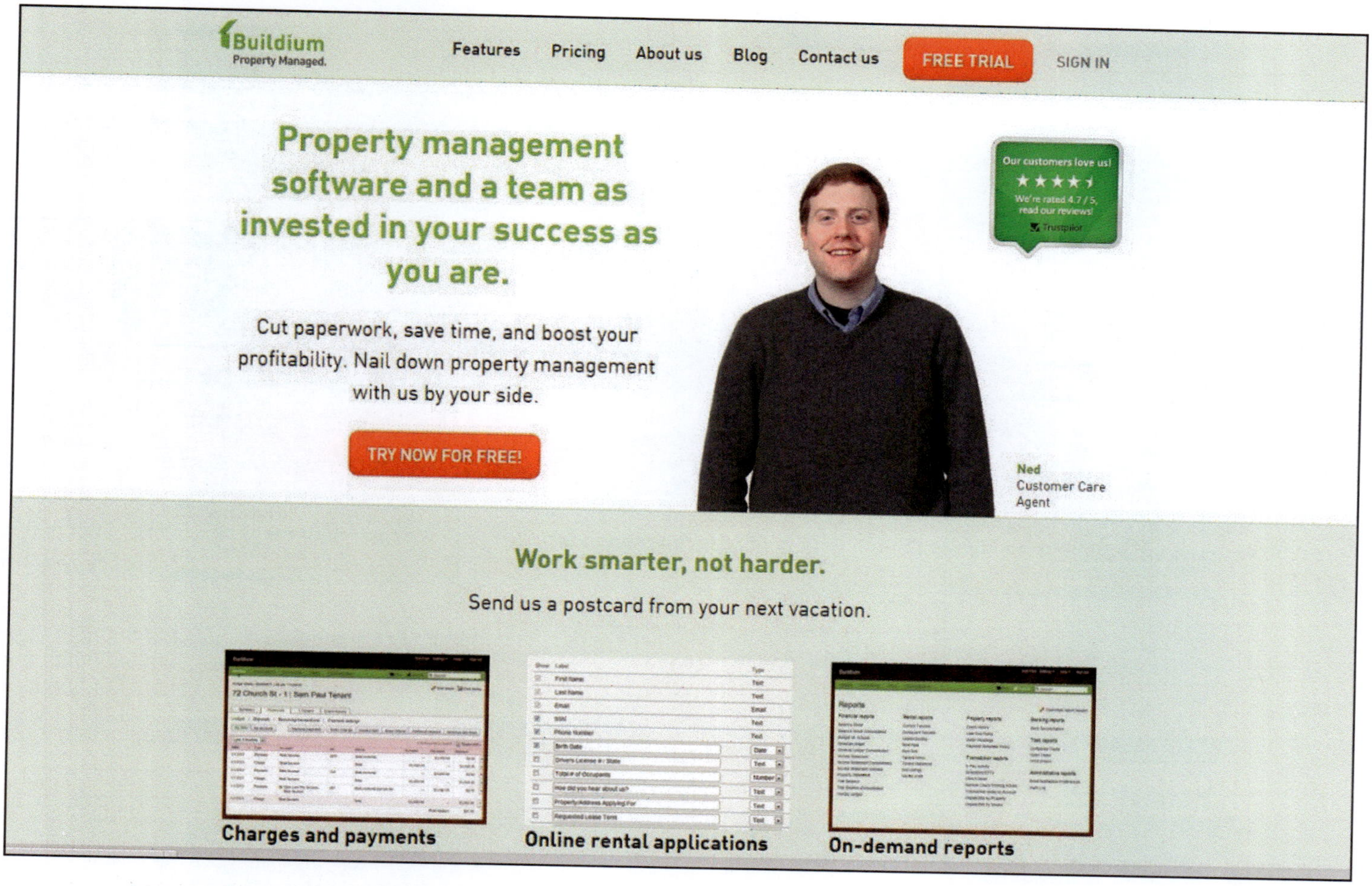

Buildium is web-based property management software that enables managers of residential and association properties to take complete control of every aspect of their business, including the rent, vacancy, and maintenance cycles.

Buildium streamlines management of the rent cycle with key features such as tenant and lease tracking, full general ledger accounting, automated rent and late fee reminders, and on-demand reports. Additionally, rent collection can be automated; property managers can receive online payments via **Buildium's** tenant portal.

Efficient management of the vacancy cycle begins with syndicating vacancies to **Buildium's** listing partners, including Zillow, Trulia, and Craigslist, as well as **Buildium's** own free marketing website. When applicants fill out **Buildium's** customizable online rental applications, the information is saved instantly to the **Buildium** tenant database. **Buildium** also offers online tenant screening services for running background checks.

14 - AppFolio.com

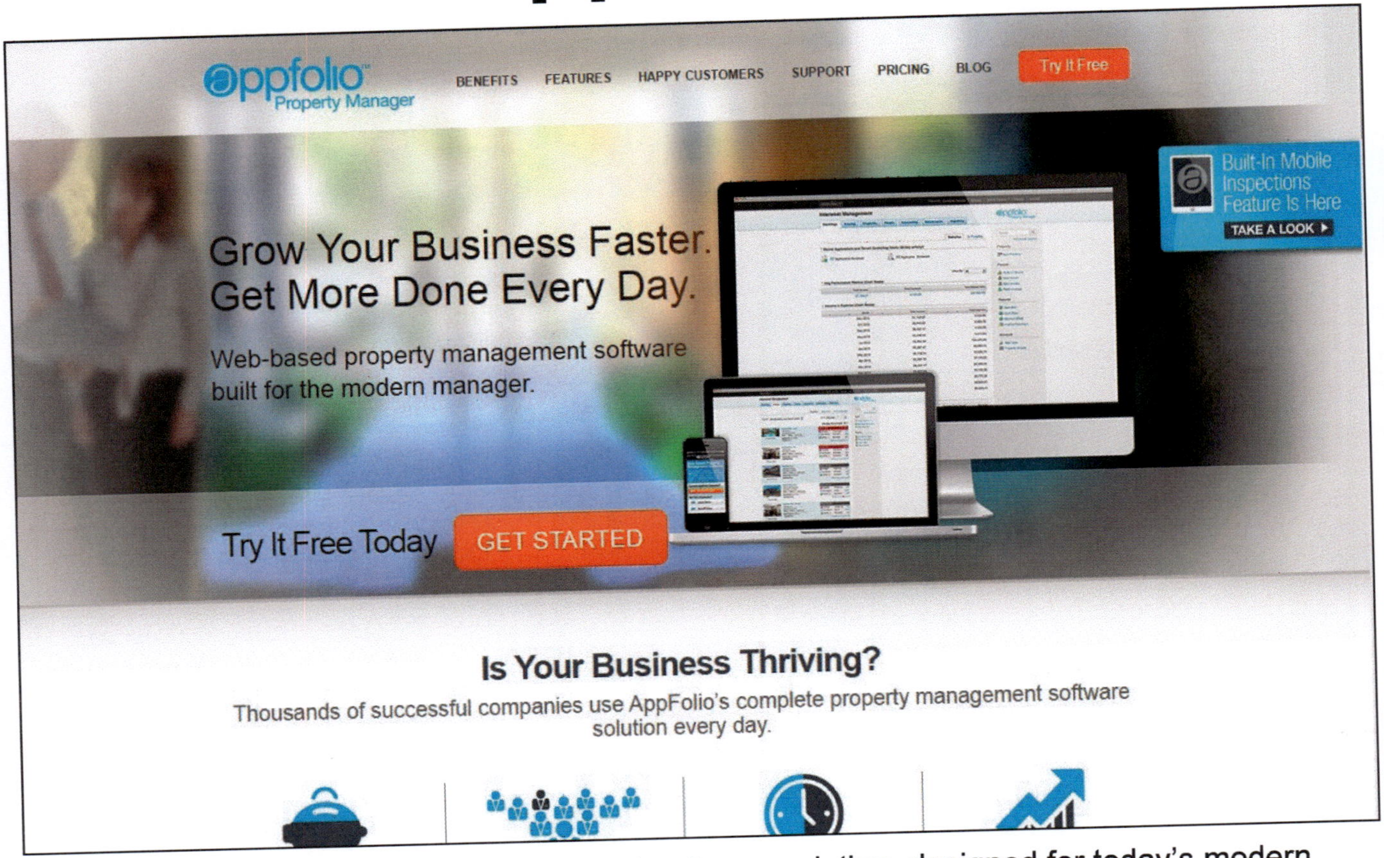

AppFolio is a complete property management software solution, designed for today's modern property manager and real estate investor. This site allows you to manage your properties from anywhere and at any computer. **Appfolio** can accommodate investors of any size - you pay a fee per listed property. Since **Appfolio** is a complete web-based property management software it makes it simple to drastically reduce the paper you use in your office.

Key features of **AppFolio**.com:

- Web based so there are no updates or downloads required.
- Unlimited users so your whole team can have personalized access.
- Training and technical support included.
- No need to upgrade or switch plans. You pay a monthly service fee plus only 1.00 per listed unit.
- Easily post vacancies to many third party sites with a few clicks in **Appfolio**.com
- Upload and store as many documents as needed to run your business - no limits.
- One monthly fee based on the number of properties you list, with no extra costs.
- Extras offered including: tenant liability insurance, a personalized website for your company, resident screening, and a 24/7 property management answering service.

15 - PropertyManagerCloud.com

Property Manager Cloud helps you successfully manage your real estate business anywhere, anytime, and on any computer. It is a totally online and cloud based way to manage your properties.

Key features:

- Completely paperless system that stores all of your information in the cloud.
- Tracking tools for income and expenses. Includes key financial reports and graphs covering every aspect of your business.
- Manage all work orders, requests and follow up online. Estimates, pictures, and details can be added for each work order.
- Includes a tenant portal where residents can make payments, apply for housing or submit maintenance requests.
- Publish vacancies to a number of external websites with ease right from the site.
- When you are logged into Property Manager Cloud, your data is encrypted and secured.
- Plans are billed monthly and you can upgrade to a larger plan at anytime.
- No software to install or upload. All updates are automatic and free.
- Free technical support via phone or email.

16 - eppraisal.com

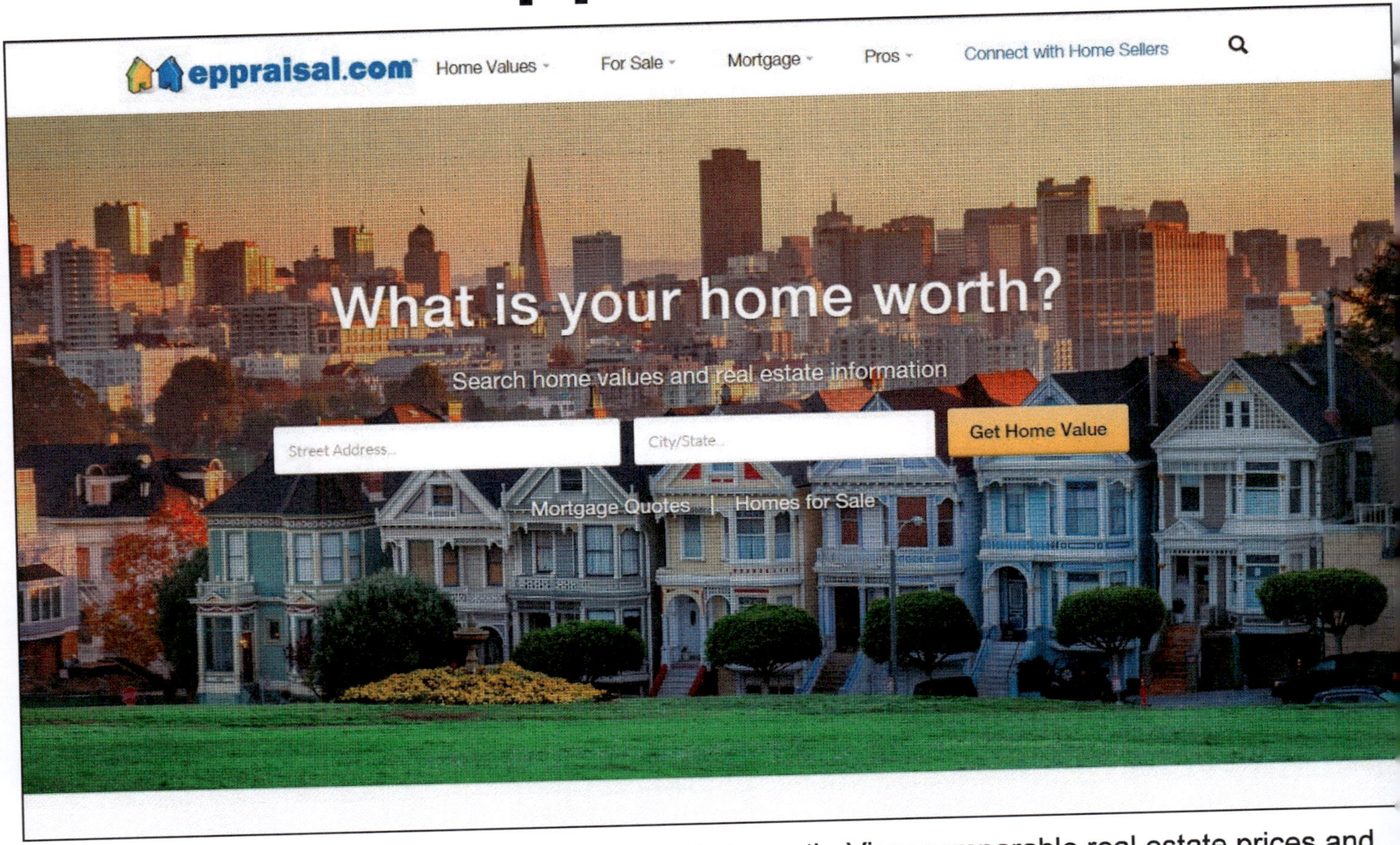

Free home valuation estimate on what your property is worth. View comparable real estate prices and recent property sales in your neighborhood.

An **eppraisal** is an estimated market value of a property, which is calculated based on recent sales of similar properties. In addition, an **eppraisal** report includes detailed property information, analysis of the surrounding area, and comparable properties.

17 - FindCompsNow.com

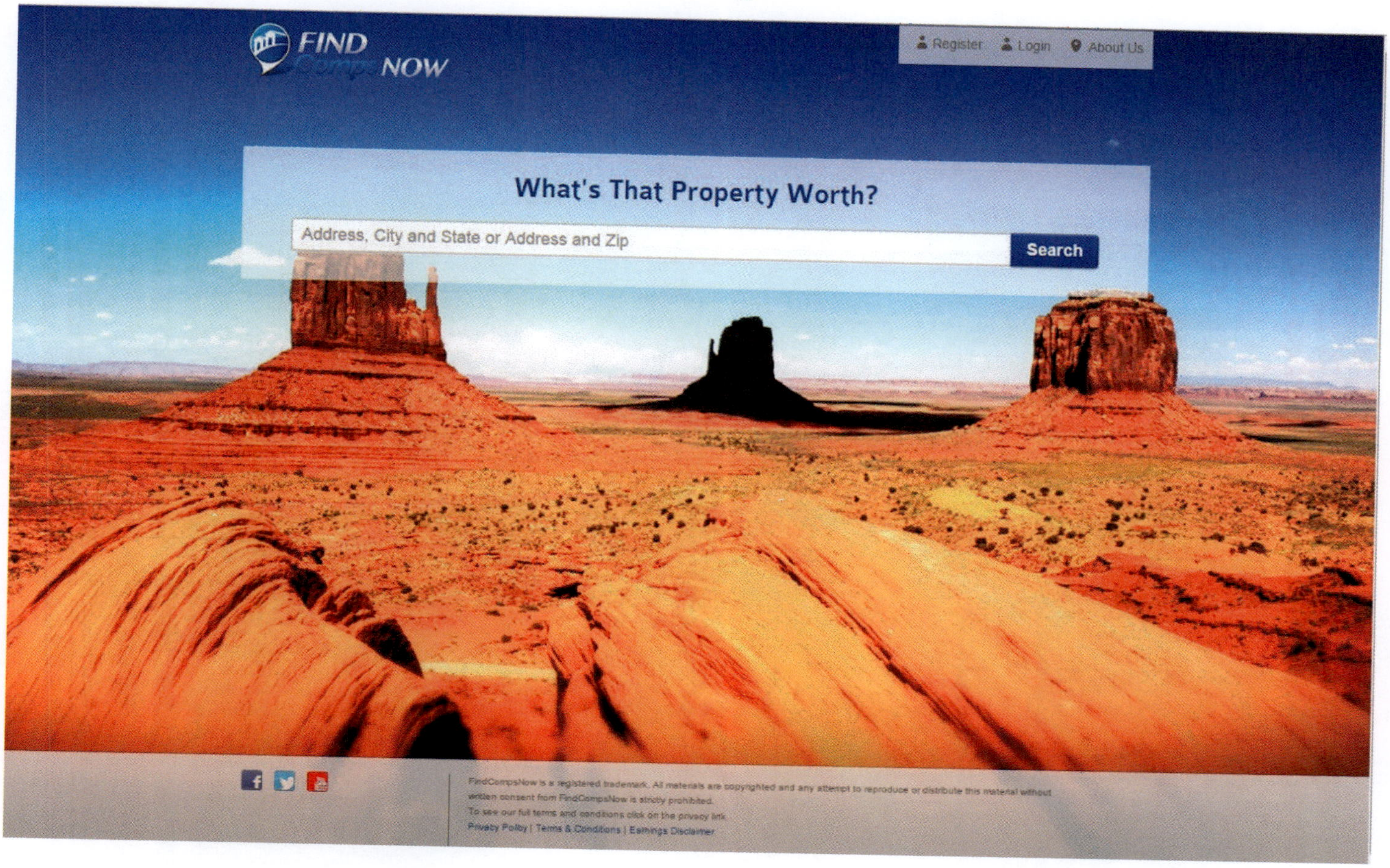

FindCompsNow.com has solved that problem for you.You now have the ability to find cash buyer comps for FREE in ALL 50 states. And check this out - all in one place!

Estimate market value of a property.

18 - VoiceConnectINC.com

VoiceConnect has been providing communication services to businesses, associations, entrepreneurs and individuals for many years. They offer a wide variety of telecommunications services to suit the needs of any organization. **VoiceConnect** pride themselves in offering top-notch services and personal customer support.

Used by Real Estate Investors for Live Answering Service, especially wholesalers.

19 - GoalsonTrack.com

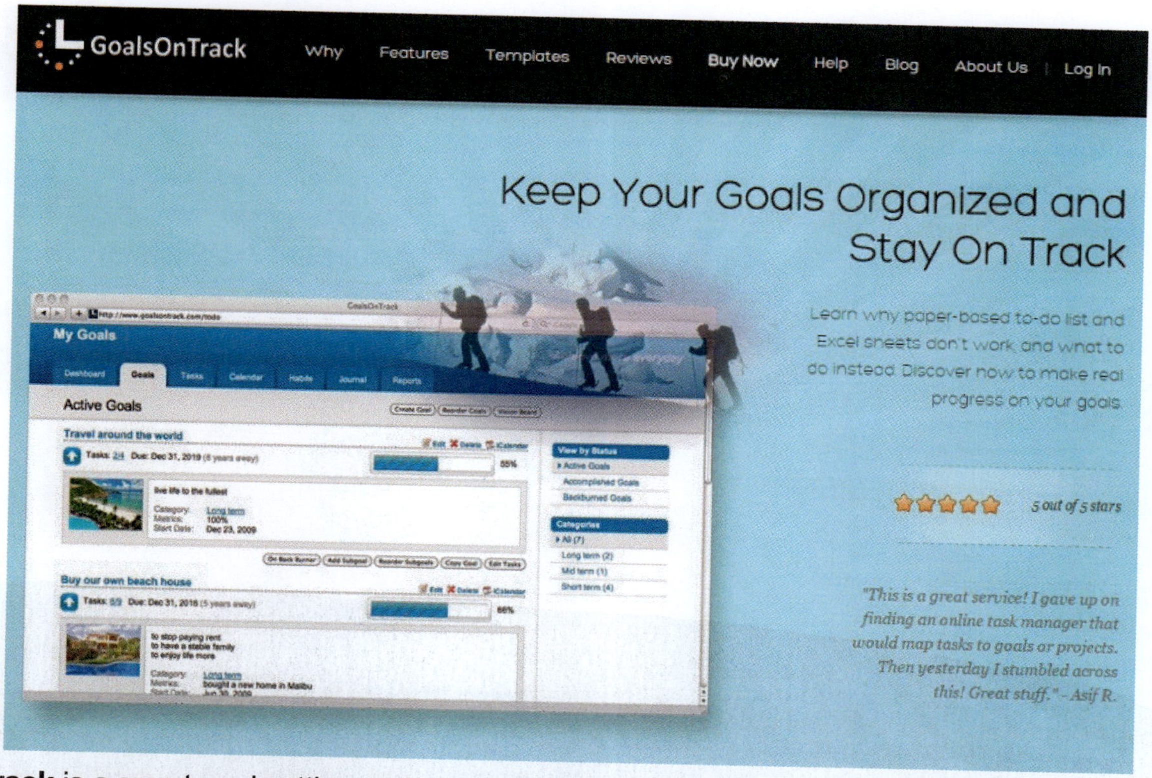

GoalsOnTrack is a smart goal setting and personal development application that helps you set and track goals. **GoalsOnTrack** can manage tasks, track time, build habits and keep a goal journal. It offers a comprehensive set of tools to guide you through a successful goal achieving process. The process is setting goals, creating action plans, tracking progress, to forming habits and keeping goal journals and finally reaching your goals.

Key features of **GoalsOnTrack**:

- Create SMART goals to ensure your goals are Specific, Measurable, Attainable, Realistic and Timely.
- Chunk it Down - break long, complex, overwhelming goals into small, manageable sub goals to chunk it down and stop procrastination
- Real time progress tracking features allow you to view instant progress on your goals whenever you complete a task.
- Setup your tasks for your goals, and see them automatically organized by goals, categories, and days
- Track how long you spend on tasks accurately and effortlessly, using animated and offline time trackers no matter where you complete the task.
- Habit Tracker tool to help you reach goals quicker.
- Tracking your habits by placing check-marks on habits tracker calendar tool.
- Automatically track your habit strength and days to form.
- Goal journal tool to write down what you did for your goals, how you did it, and what lessons you learned.

It allows you to record the goal, start date, end date, metrics, subgoals, habits, and action plans.
It also syncs due dates with Google Calendar and Outlook. You can even create goal templates.

20 - Weebly.com

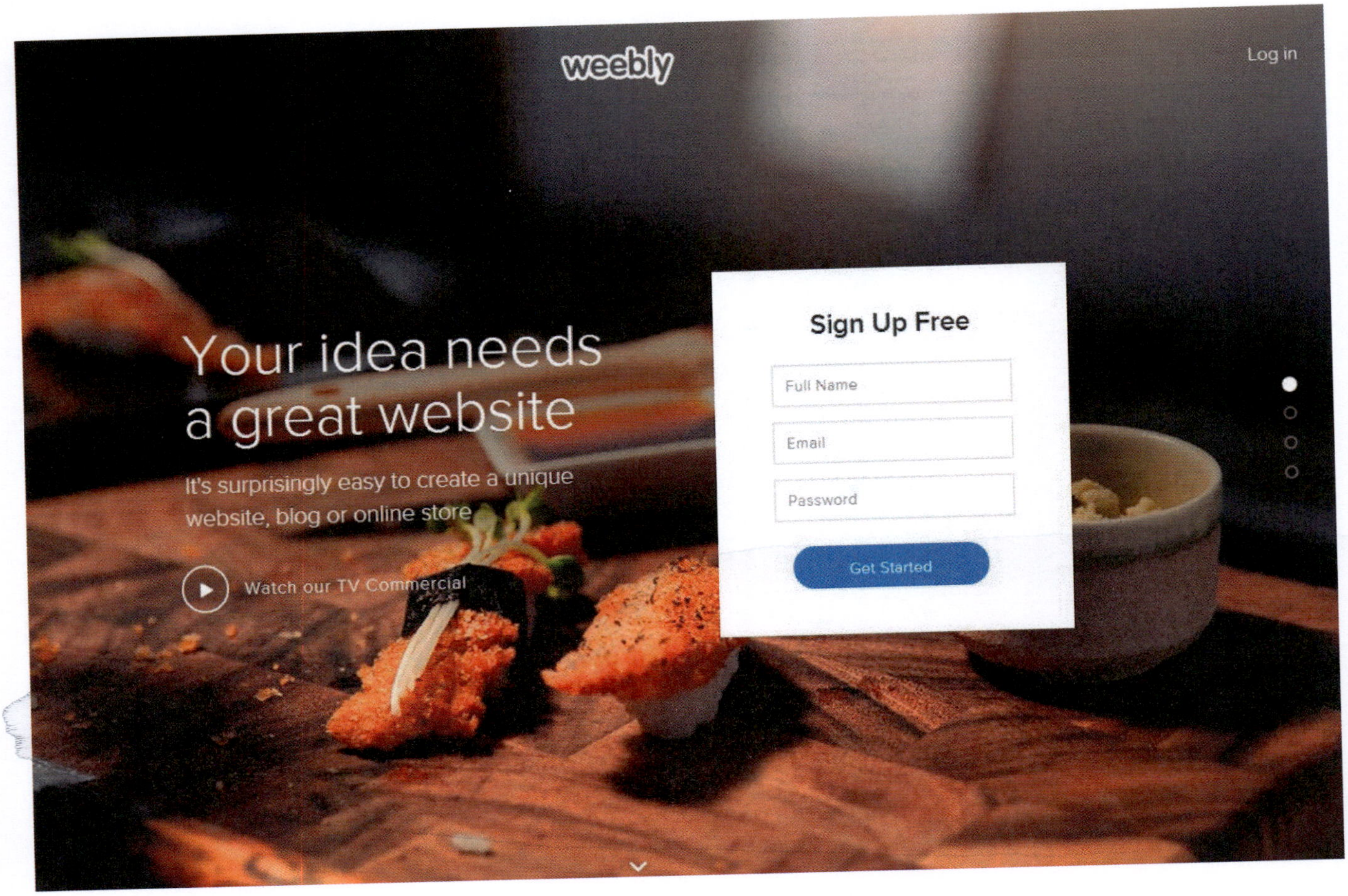

Weebly is a web-hosting service that allows the user to "drag-and-drop" while using their website builder. As of August 2012, **Weebly** hosts over 20 million sites with a monthly rate of over 1 million unique visitors.

Weebly gives millions of users a surprisingly easy and affordable ways to create a site that is as unique as they are. With a **Weebly** site, people can start their own business, communicate with their clients, showcase their achievements, and be an authority on personal and professional interests. **Weebly** gives everyone the freedom to start a site, blog or online store that works brilliantly across computers, phones and tablets. **Weebly** offers a range of pricing options including a free plan and premium plans starting at $4/month, **Weebly** has everything you need to plan, build, publish and grow a site that meets your goals.

21 - Fiverr.com

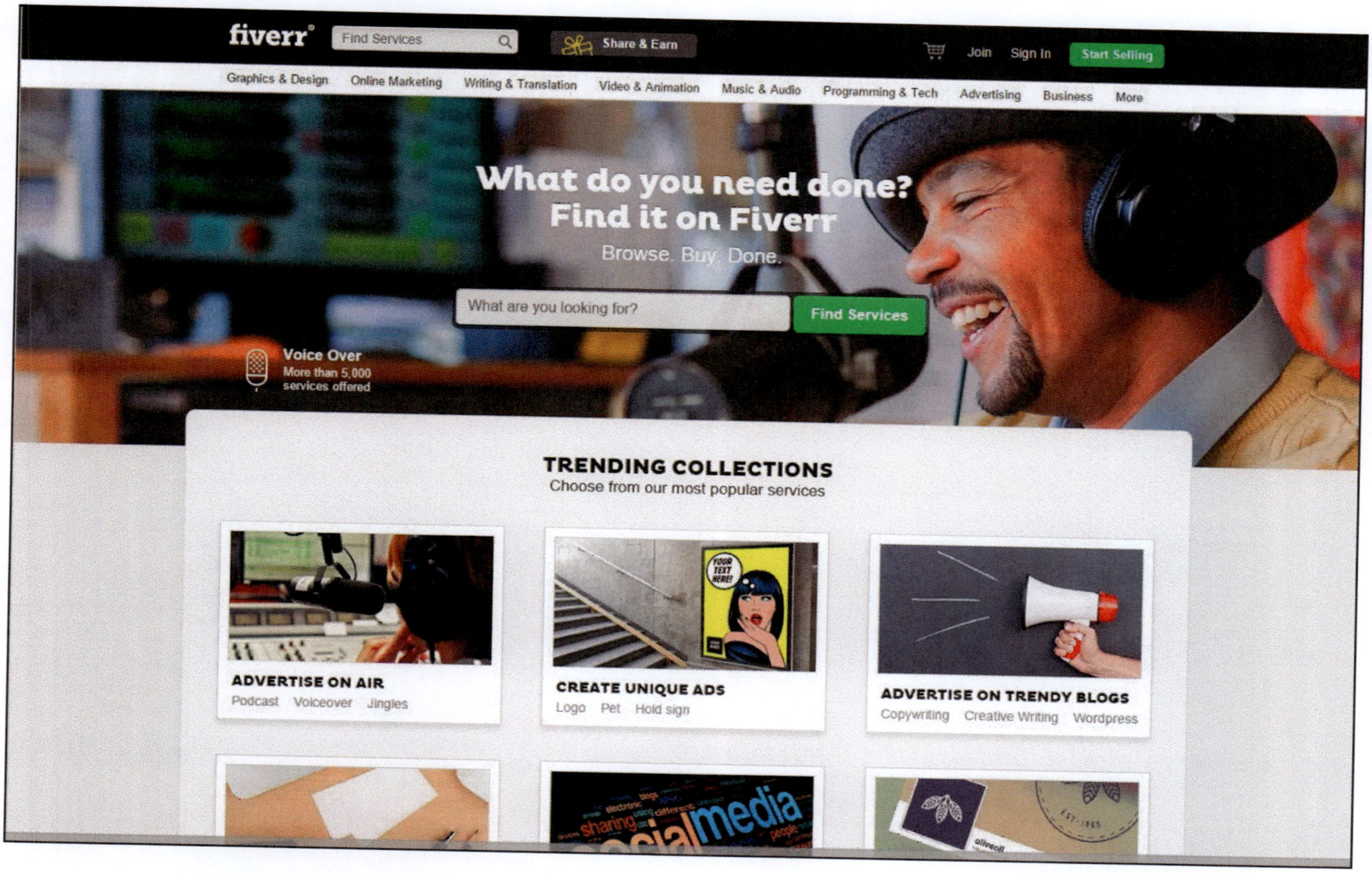

Fiverr is a global online marketplace offering tasks and services beginning at a cost of $5 per job performed.That's how it received it's name. This site is primarily used by freelancers who use **Fiverr** to offer a variety of different services to customers worldwide.

Uses:
Logo designs
Voice overs
Ideas
Data entries
Social sedia posts
Flyer designs

savvy INVESTORS .com

22 - MySmartMove.com

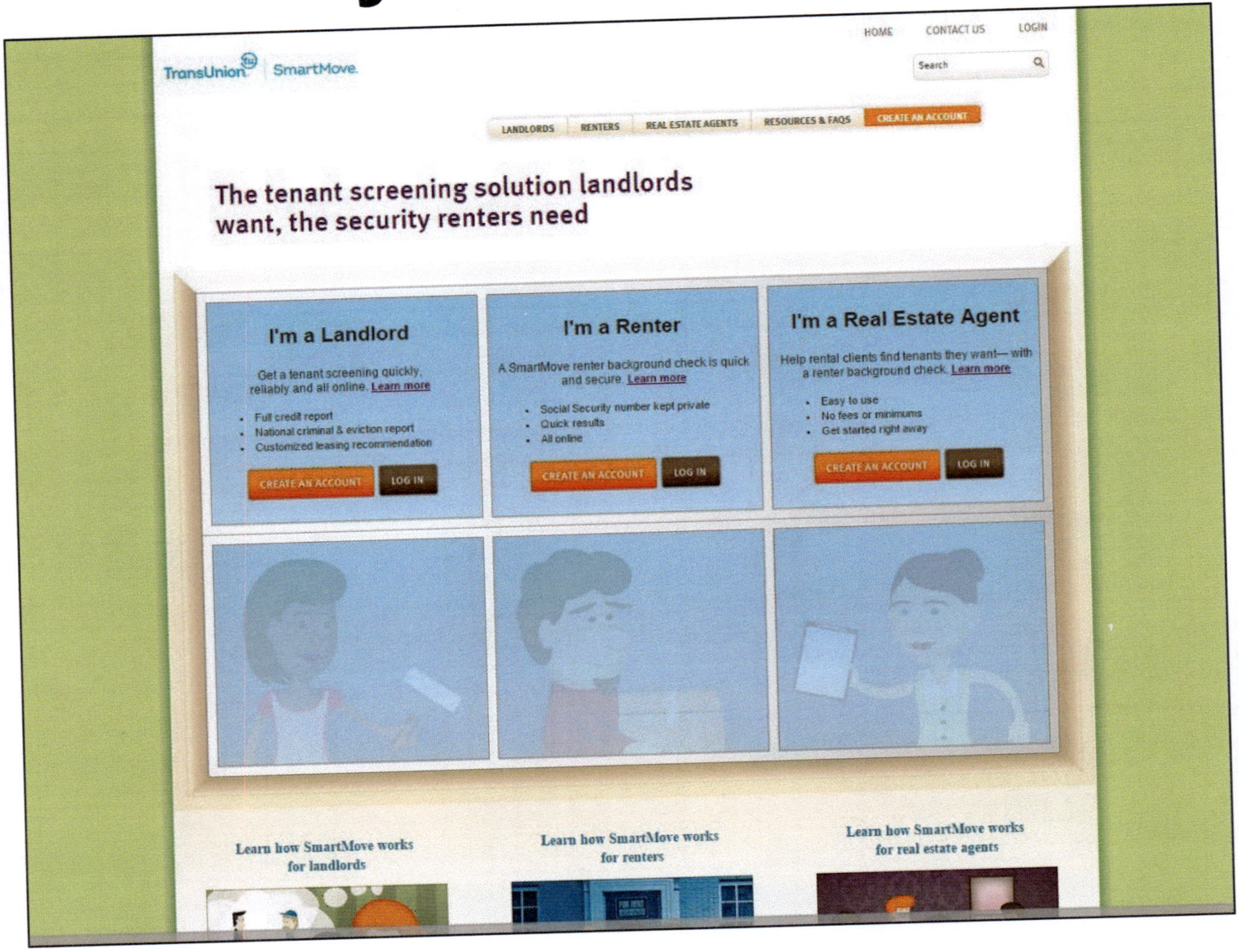

SmartMove is an online screening service for independent landlords and their rental applicants. It was designed to help smaller landlords make educated leasing decisions. **SmartMove** gives landlords access to the same tools for renter screenings that larger property management companies use.

23 - PatLive.com

Patlive Receptionist service has all the features you need at an affordable price. There are no hidden charges, and there are no contracts for you to sign.

Their Services: 24/7
Live Answering Service
Call Center Service
Local Numbers
Toll Free Service
Fax Service
Call Forwarding
Email To Fax
Voice Mail

24 - AirBnB.com

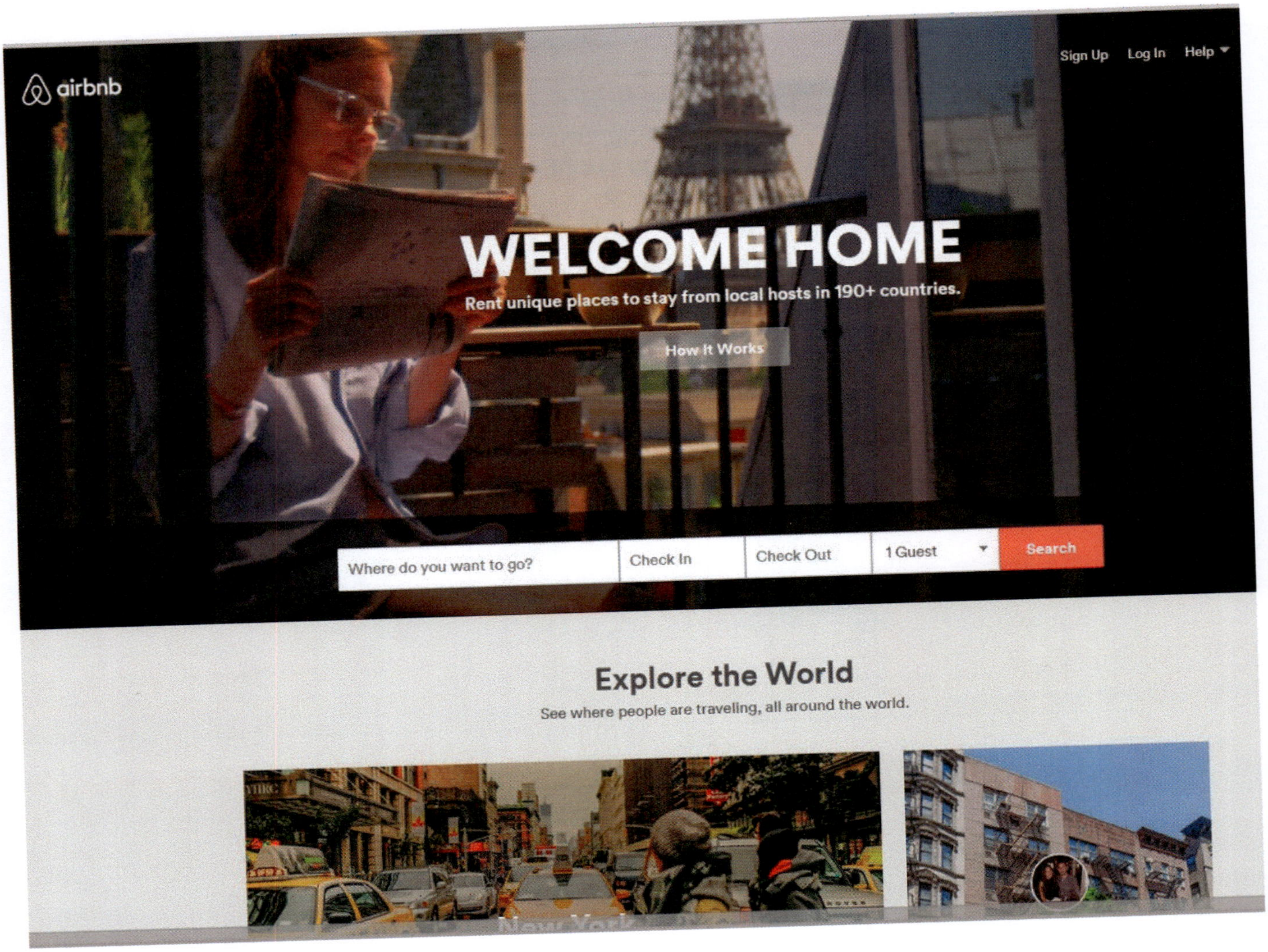

Airbnb is an online community marketplace that connects people looking to rent their homes with people who are looking for accommodations. **Airbnb** users include hosts and travelers. Hosts list and rent out their unused spaces. Travelers search for and book accommodations in 192 countries worldwide.

25 - BookBub.com

BookBub is a service that promotes limited-time discounts on ebooks. Readers can sign up to receive a daily alerts with deals in categories that match their interests. Authors, publishers, and agents partner with **BookBub** to reach the millions of book lovers who use their service.

26 - GoSection8.com

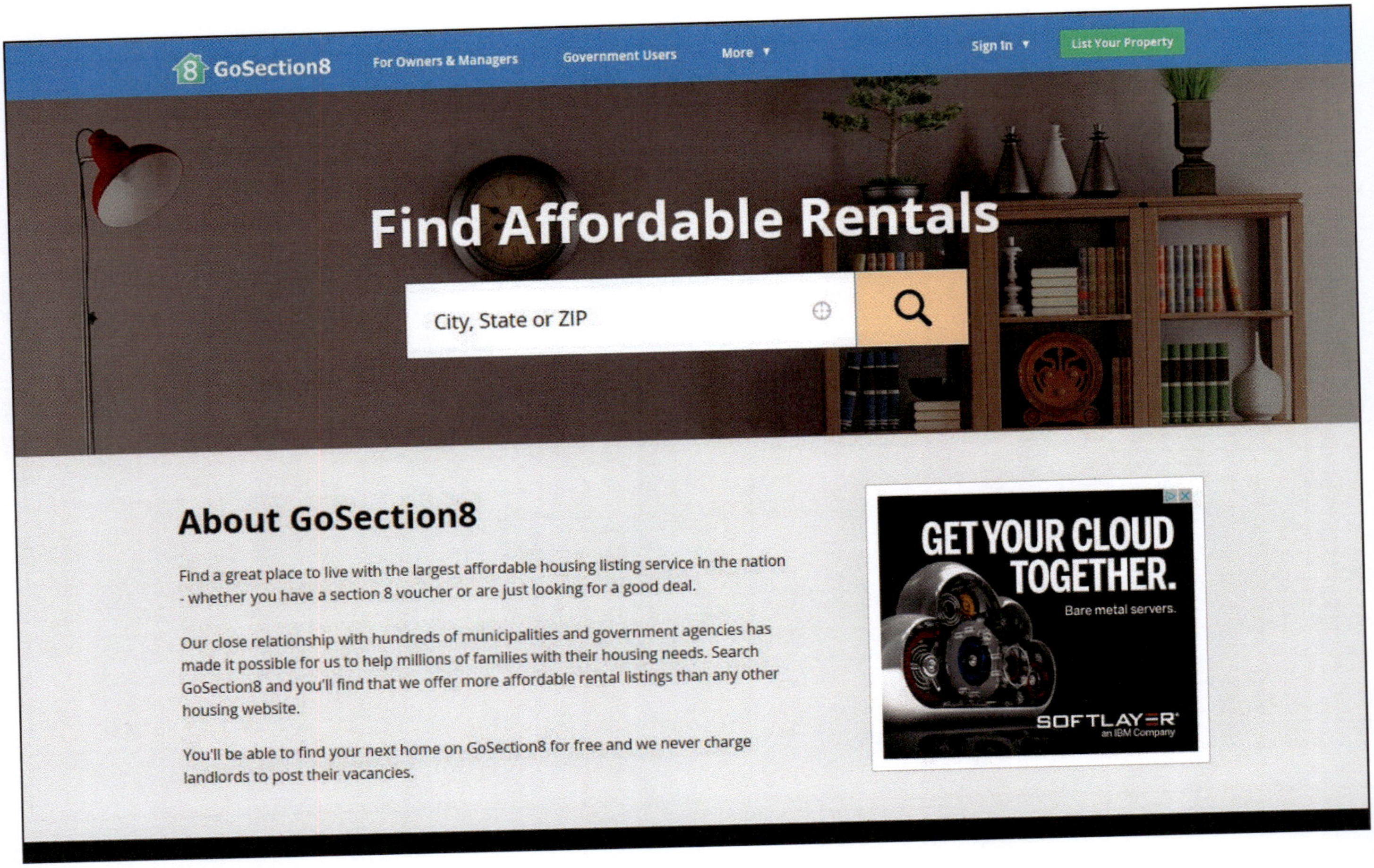

GoSection8.com works with landlords and public housing agencies to maintain the largest and most accurate listings of voucher program rentals available in every market in the U.S. Tenants are not charged for receiving the listings, which are available to them online, via e-mail alerts, by calling our toll-free number or from their local housing authority. **GoSection8's** goal is to connect landlords with tenants and assist the public housing agencies by maintaining the most accurate, comprehensive listings of certified rentals in their market.

savvy INVESTORS .com

27 - ACWholesalers.com

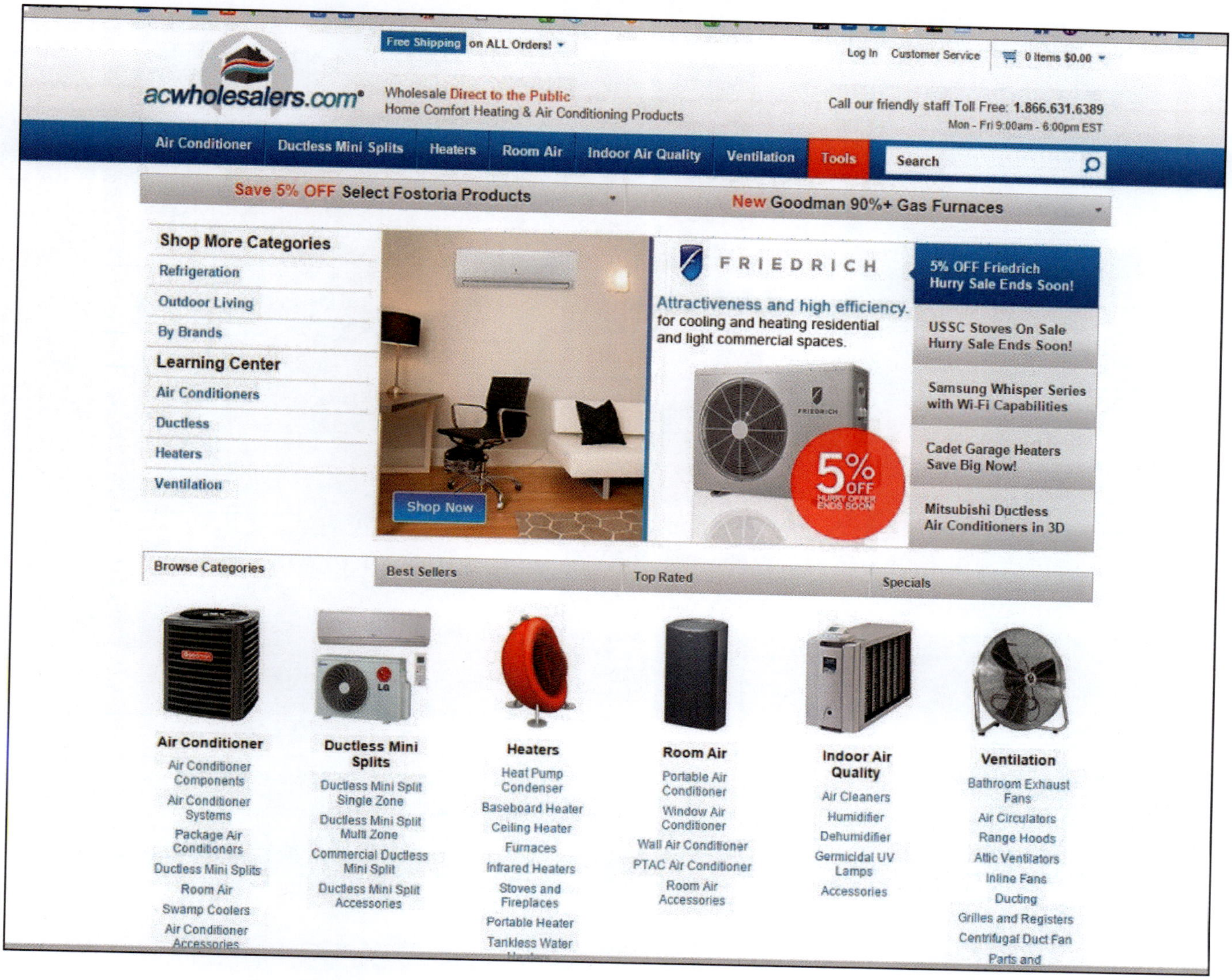

AC Wholesalers sells Goodman heat pump quality air conditioning, Mitsubishi air conditioners, and heating and indoor air-quality products wholesale directly to you.

28 - MetroFax.com

MetroFax enables customers to easily access faxes from their laptop, desktop or wireless device. Faxes are delivered as attachments directly to the customer's email inbox. Individuals and companies of all sizes depend on **MetroFax's** simple, affordable, and scalable Internet fax service. It includes a choice of inbound toll-free or local fax numbers in thousands of cities across the U.S; and with no extra charges for long distance or transmission time charges

MetroFax is a leading provider of Internet faxing with advanced features and flexible usage plans. Powered by the largest digital fax network in the world, and equipped with cutting-edge servers that provide unparalleled transmission security and maintain 99.9% up-time.

29 - DropBox.com

Dropbox is a home for all your photos, docs, videos, and files. Anything you add to **Dropbox** will automatically show up on all your computers, phones and even the **Dropbox** website — so you can access your stuff from anywhere.

30 - PassPack.com

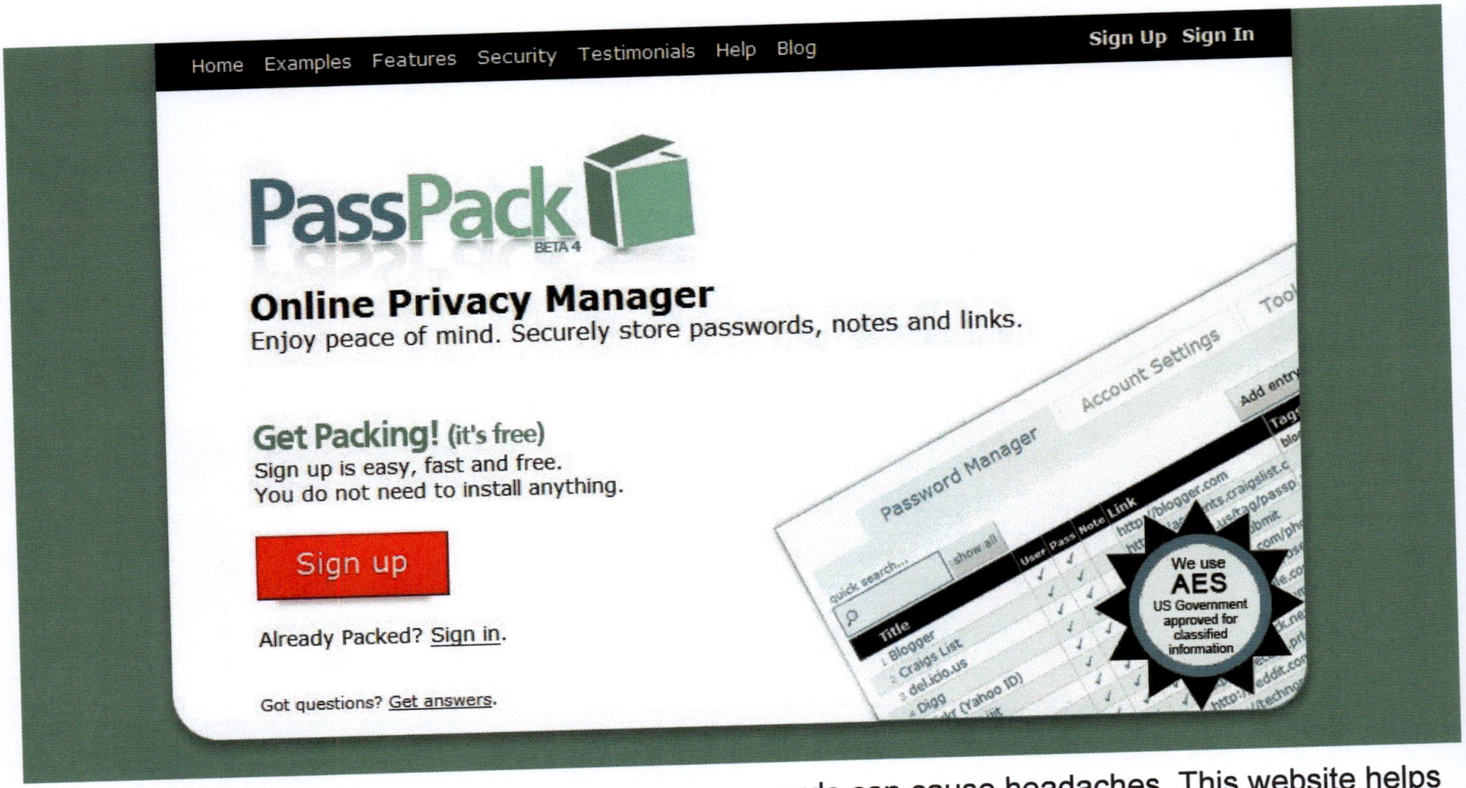

Strong passwords combined with having many passwords can cause headaches. This website helps eliminate that by letting you tag, sort, search and manage multiple logins per site.

Collaborative: Family, co-workers and team members all need access to shared accounts. Securely share and provision passwords using military grade encryption.

Host Proof: They never have your un-encrypted data. Even if our servers were hacked your passwords would be safe.

Everywhere: **Passpack** is available to you from any browser, anywhere, anytime, no need to carry yet another device or install on multiple computers.

1 Click. Simple:Your personal login button. Just install it once in any browser, then click to login to your favorite websites.

Customers from Free Accounts to Enterprise managers have a great experience with Passpack.

For free for holding 100 passwords. Paid accounts you can share and keep 1,500.

savvy INVESTORS .com

31 - withoomph.com

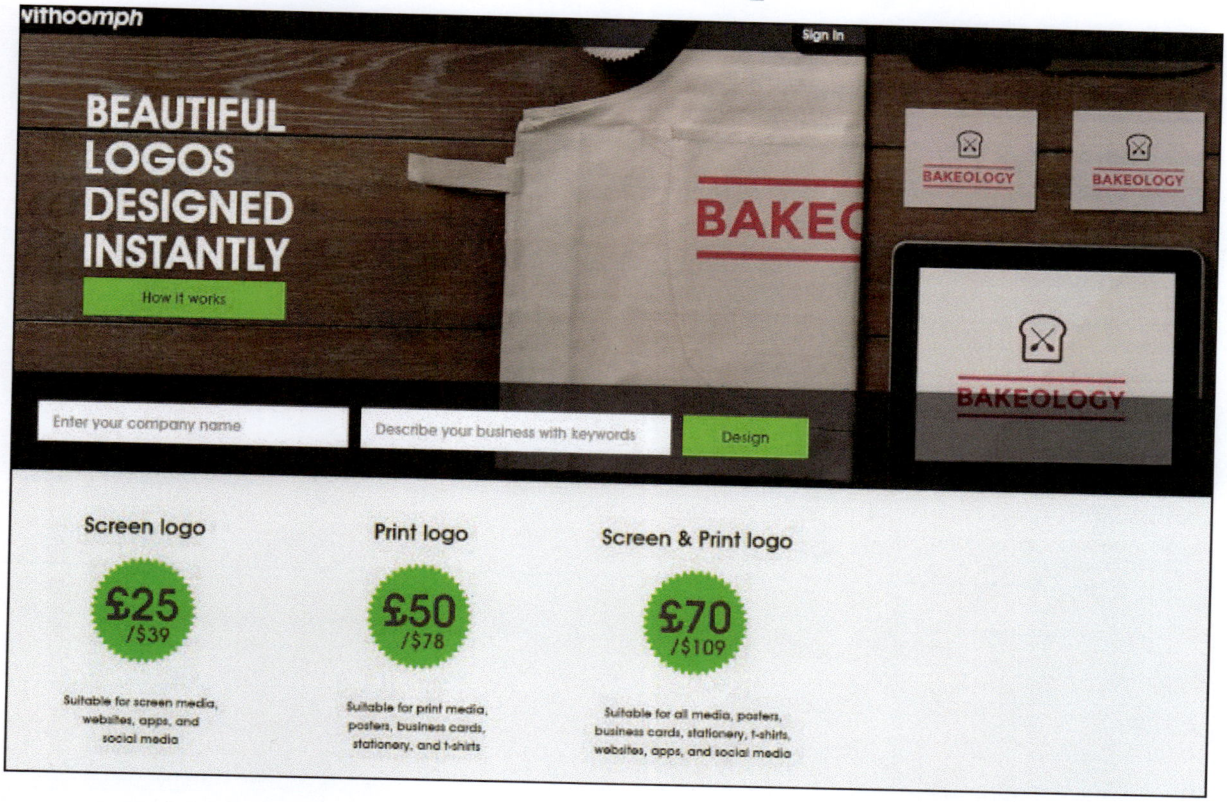

Get fast business logo ideas with **Withoomph** for free!!

32 - google.com/drive

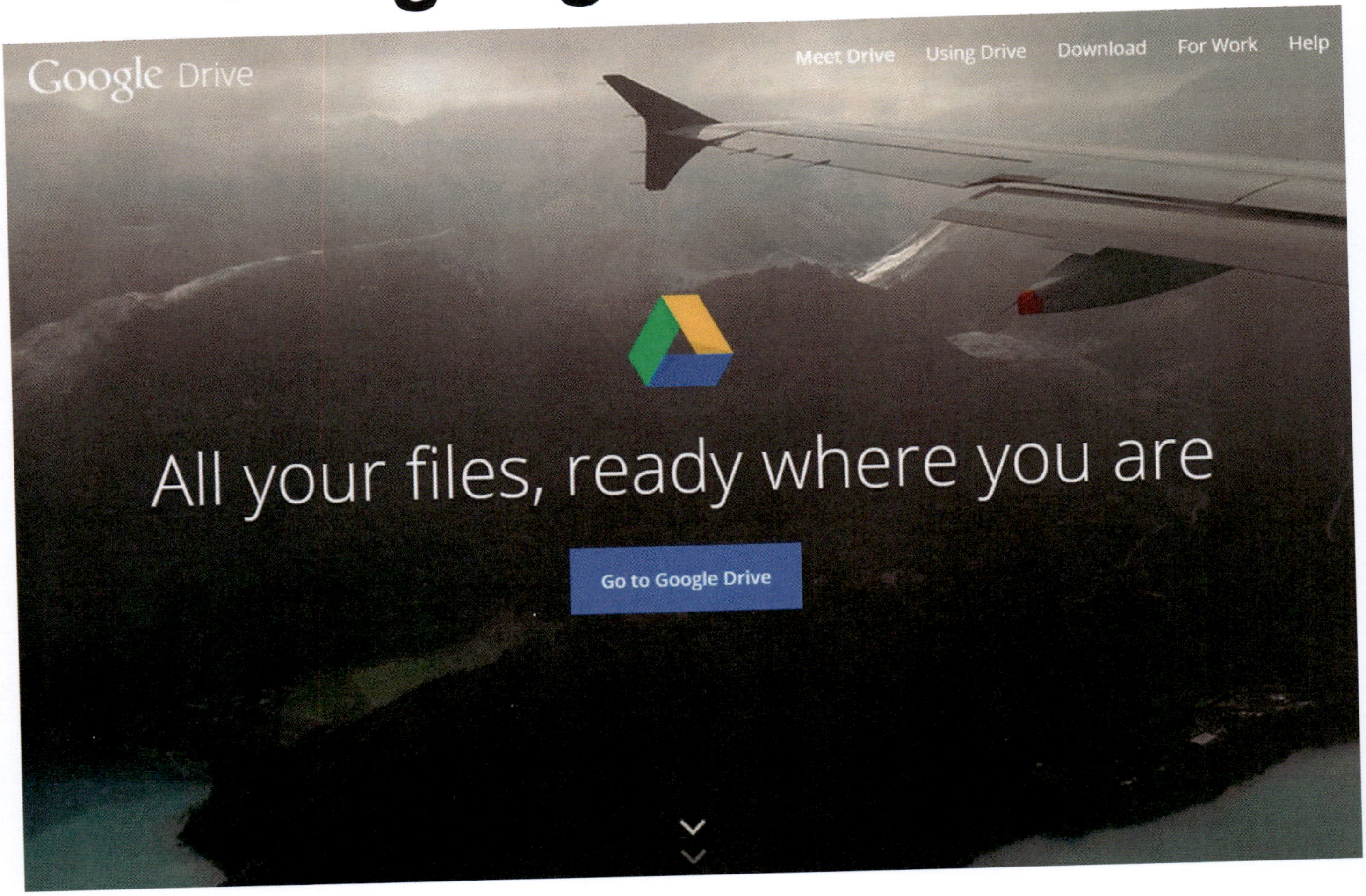

Google Drive is a file storage and synchronization service created and managed by **Google**. It allows users to store documents in the cloud, share files, and edit documents with collaborators.

Use **Google Drive** on the web to store and organize all your files. You get 15 GB of free storage across **Google Drive**, Gmail, and Google+ Photos. If you run out, you can buy more storage. With **Google Drive** on the web, you can: Create, add, or upload a file with a single button.

Get access to files anywhere through secure cloud storage and file backup for your photos, videos, files and more with **Google Drive**.

33 - ClearNow.com

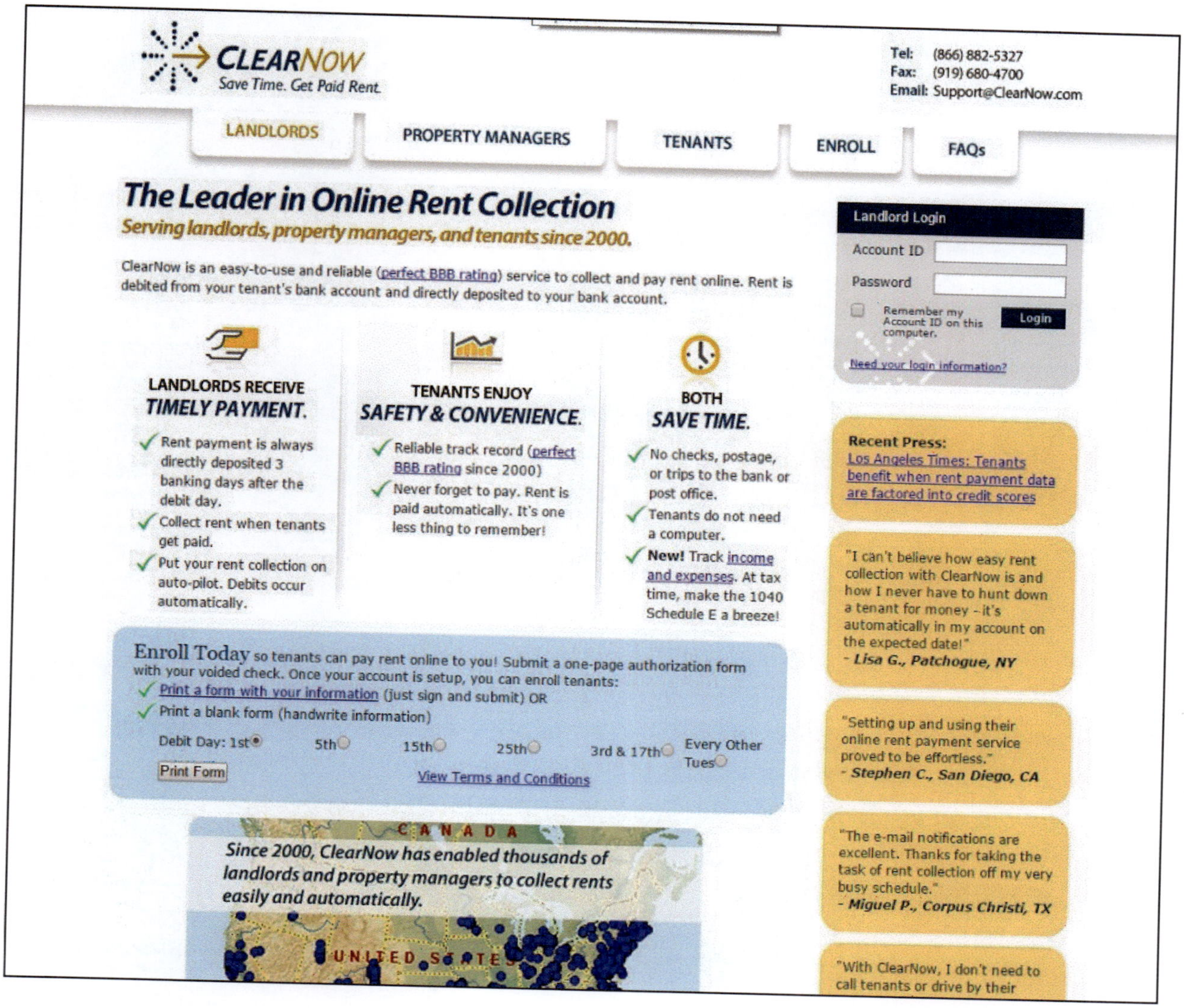

ClearNow is the easiest, most cost-effective way to collect rents from tenants. Rents are automatically withdrawn from tenant bank accounts and directly deposited into your bank account. **ClearNow** saves you time, eliminates headaches, and empowers tenants to pay you!

- Automatically receive rent - don't wait for checks or make trips to the bank.
- Get paid on time - deposits are always three banking days after the debit day.
- Collect when tenants get paid - biweekly, semimonthly, or monthly.

ClearNow has been in business since 2000, serving thousands of landlords, property managers, and their tenants:

- You will receive prompt notification of who has paid.
- Tenants do not need a computer to use **ClearNow**.

34 - Auction.com

Auction.com is a real estate marketplace where you can buy and sell residential and commerical properties in an online auction. Stay on top of the real estate that interests you by searching, saving, registering and bidding. **Auction.com** hosts online auctions for properties all across the USA - all you need to find and bid on these properties is a computer!

Key features of **Auction.com**:

- Search and bid on properties anywhere in the country at anytime from your computer, smart phone, or tablet.
- They post a variety of commercial and residential properties: foreclosures, short sales, FSBO, and listings for local sheriffs sales.
- Filter searches by location, property type and asset type.
- Save and name your favorite properties for easy access later on.
- Bidding is real time and auction.com will send you an alert e-mail to let you know if you've been outbid.
- No realtor fees. Some properties have a buyers premium fee, but some do not! Check the listing you are interested in.
- If you win a property you are assigned a specialist with a direct phone and email to help answer questions.
- Keep entity documents on file with Auction.com after a first purchase to make future purchases easier.
- There is an available app for Apple and Android devices.

This website is basically an ebay for houses. The online tools make buying a property on an auction simple.

savvy INVESTORS .com

35 - Evernote.com

Evernote is a suite type of software and services, designed for notetaking and archiving. A "note" can be a piece of formatted text, a full webpage or webpage excerpt, a photograph, a voice memo, or a handwritten "ink" note. Notes can also have file attachments. Notes can be sorted into folders, then tagged, annotated, edited, given comments, searched and exported as part of a notebook.

Uses:
Punch Lists, Budget Per Project, Saving Receipts and, Ideas
Screenshots of Assessors Records For Research, Journal, Saving Paint Colors and, Notes on Workers. Google for your mind!

36 - HomeWyse.com

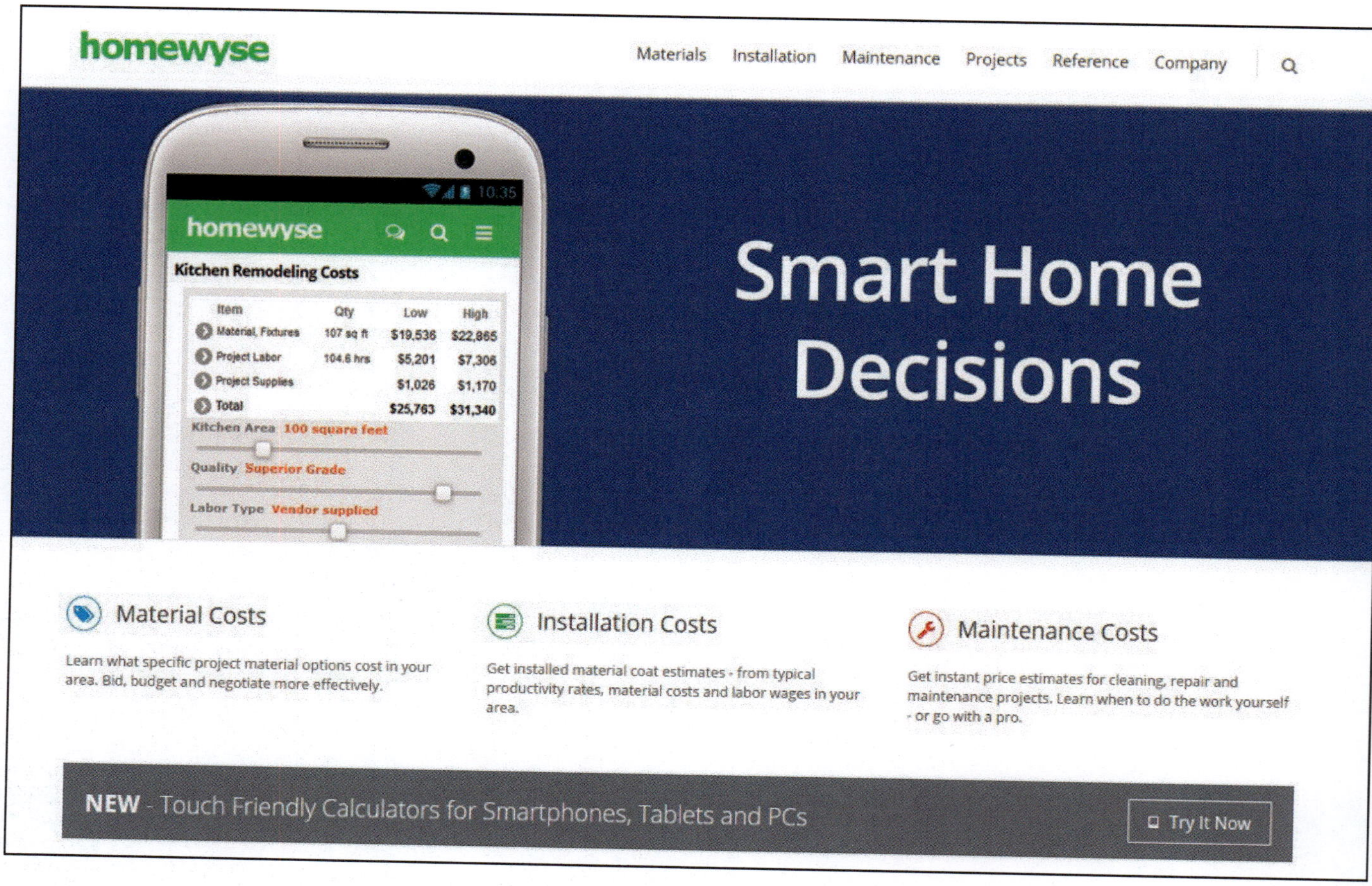

Homewyse is the complete, independent reference for home product, installation and service decisions.

Homewyse it is a site that gives you average costs for home improvement projects for your area.

37 - CrimCheck.com

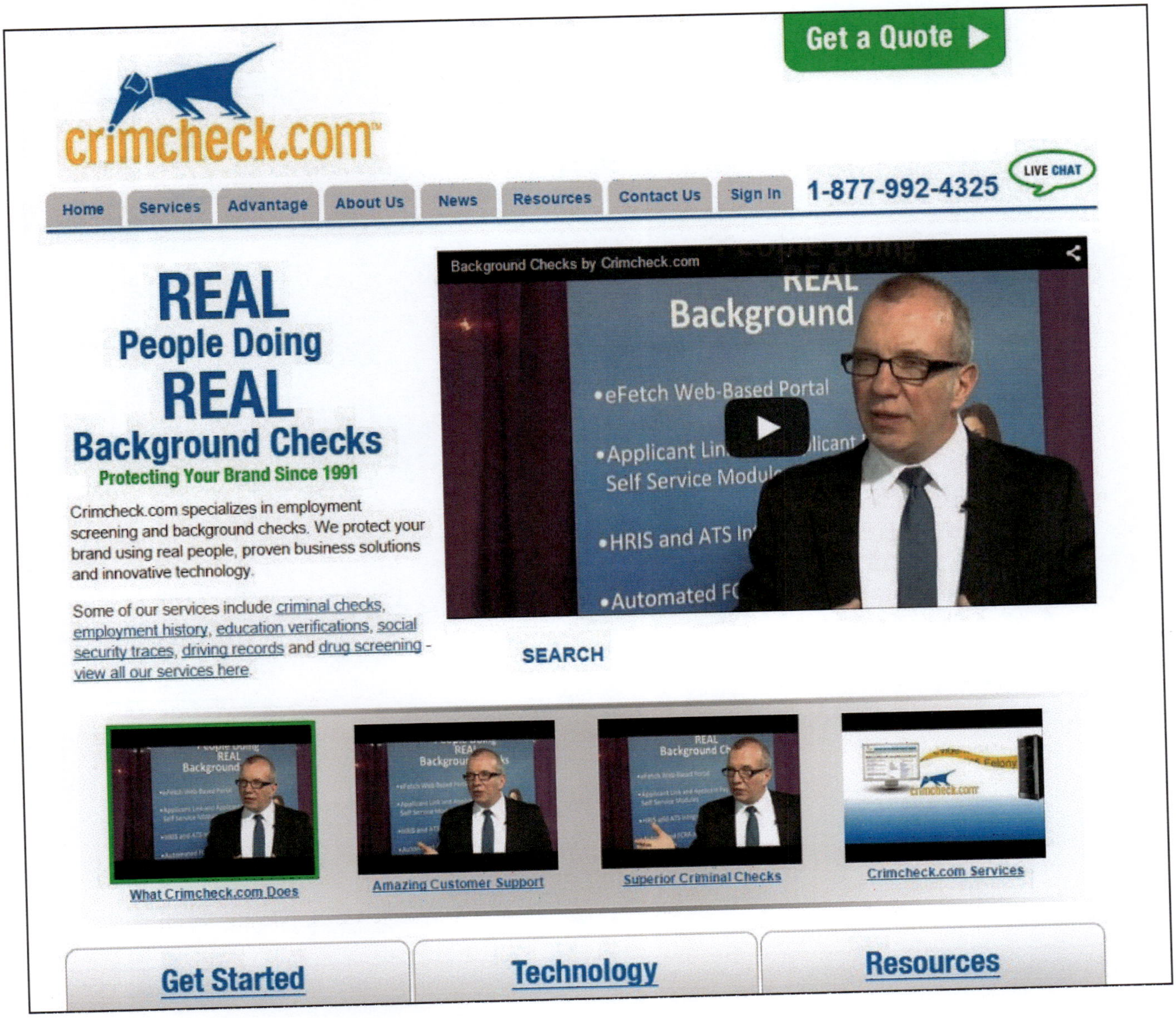

Crimcheck.com specializes in employment screening and background checks.

38 - join.me

Join.Me is an online tool that offers free screen sharing and online meetings

Free, simple and, fast online meetings. Share your screen instantly with anyone.

39 - OhDontForget.com

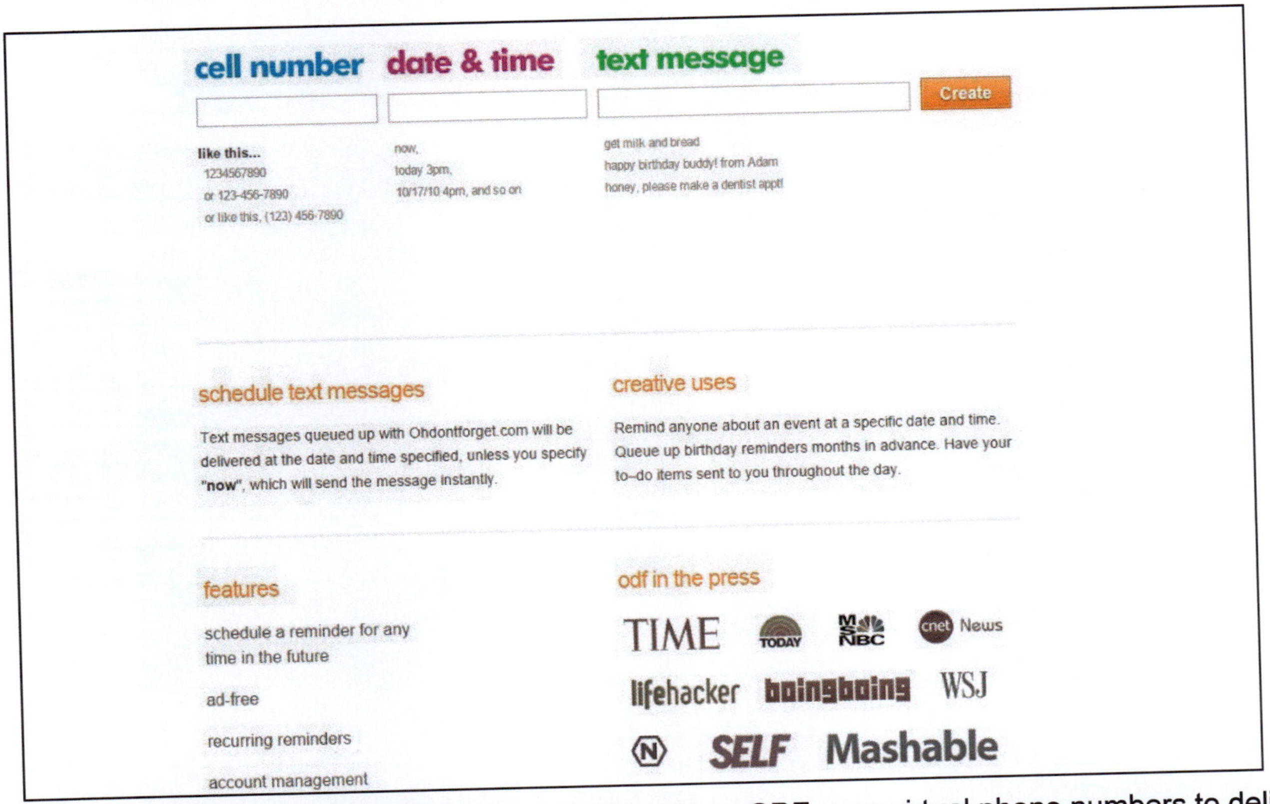

Oh Don't Forget (ODF) is an internet based reminder system. ODF uses virtual phone numbers to deliver text message reminders to you or others. Remind yourself or anyone else about an event at a specific day and time. Messages cost $0.02 to send or if you opt to deposit a larger amount into your account you can get the reduced rate of $0.01 per message.

Key features of OhDontForget.com:

- Schedule reminders for the future.
- Input anything you'd like a reminder for: meetings, shopping lists, birthdays, anniversaries, closing dates, personal or business goals, etc
- ODF saves your contacts for ease in future messaging
- Set reminders to repeat never, daily, weekly, or monthly.
- Read and view upcoming reminders before they go out.
- View reminders that have already been sent out.
- ODF blocks mass messages and marketing.
- No fees (other than what your carrier charges you) for others to receive or reply to texts from ODF.
- Users can opt to be removed from the ODF database.

Never forget meetings, closings, birthdays, or other events again. Schedule messages for right away or far in advance. Send yourself a simple list of to-do items. Send your handyman or other office staff to do or shopping lists. There are many ways this could help you build your business.

40 - CreditKarma.com

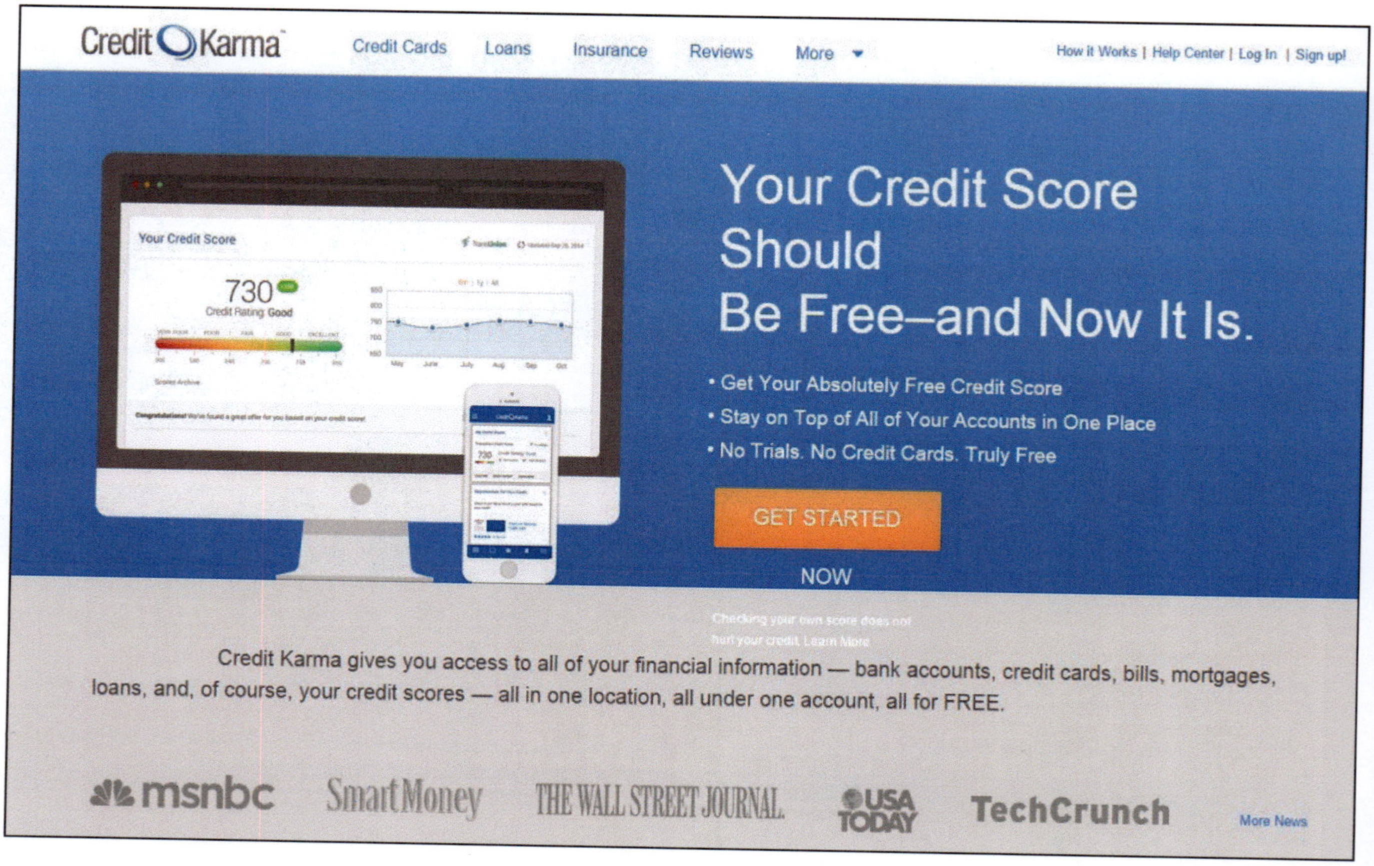

Credit Karma is a completely free pro-consumer service dedicated to demystifying the credit landscape. Their goal is that through credit simulators, free credit scores, credit advice, and credit score comparisons they can empower consumers to more actively manage their credit and their financial health. **Credit Karma** gives the customer comprehensive reports on debt, payments, accounts, inquiries, and credit history as well as the tools to improve each category.

- Totally free comprehensive credit report.
- Tools to track your credit score.
- Quickly track your credit attributes and debt over time.
- Tools to guide you towards changes that could increase your score over time.
- Notifications when new accounts or inquiries are made in your name.
- Tools and simulators to help you learn how to anticipate changes in your credit score.
- See how various items in your history affect your credit score.
- See how creditors and financial institutions see you.
- There is an available app for apple and android devices.

Your credit is an integral piece in being able to build your real estate empire. **Credit Karma** helps you keep on top of your finances, learn to improve your score over time, see how a lender will view you, and gives you an opportunity to check for fraudulent activity on your report.

41 - chrome.google.com/webstore/category/apps

The **Chrome** Web **Store** is Google's online store for web applications for Google **Chrome** or Google Apps. The software allows users to install and run web applications for the Google **Chrome** browser.

savvy INVESTORS .com

42 - google.com/voice

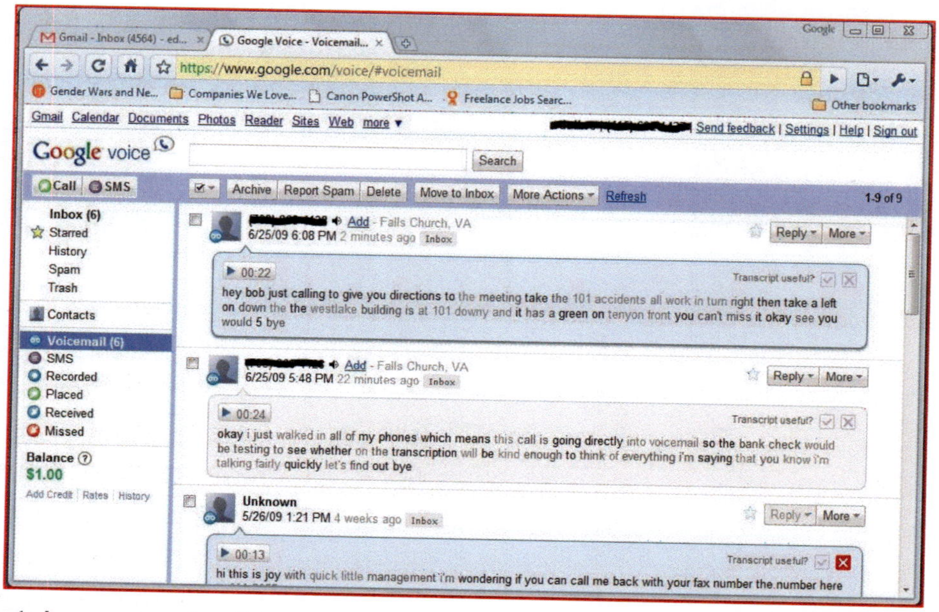

Google Voice is a telecommunications service by **Google** launched on March 11, 2009, after acquiring the service GrandCentral. The service provides a U.S. phone number, chosen by the user from available numbers in selected area codes, free of charge to each user gmail account.

Features:

* Voicemail transcription
* International calling
* Share voicemails
* Screen callers
* Personalized greetings
* SMS to email
* Block callers
* Number porting

43 - pdf2jpg.net

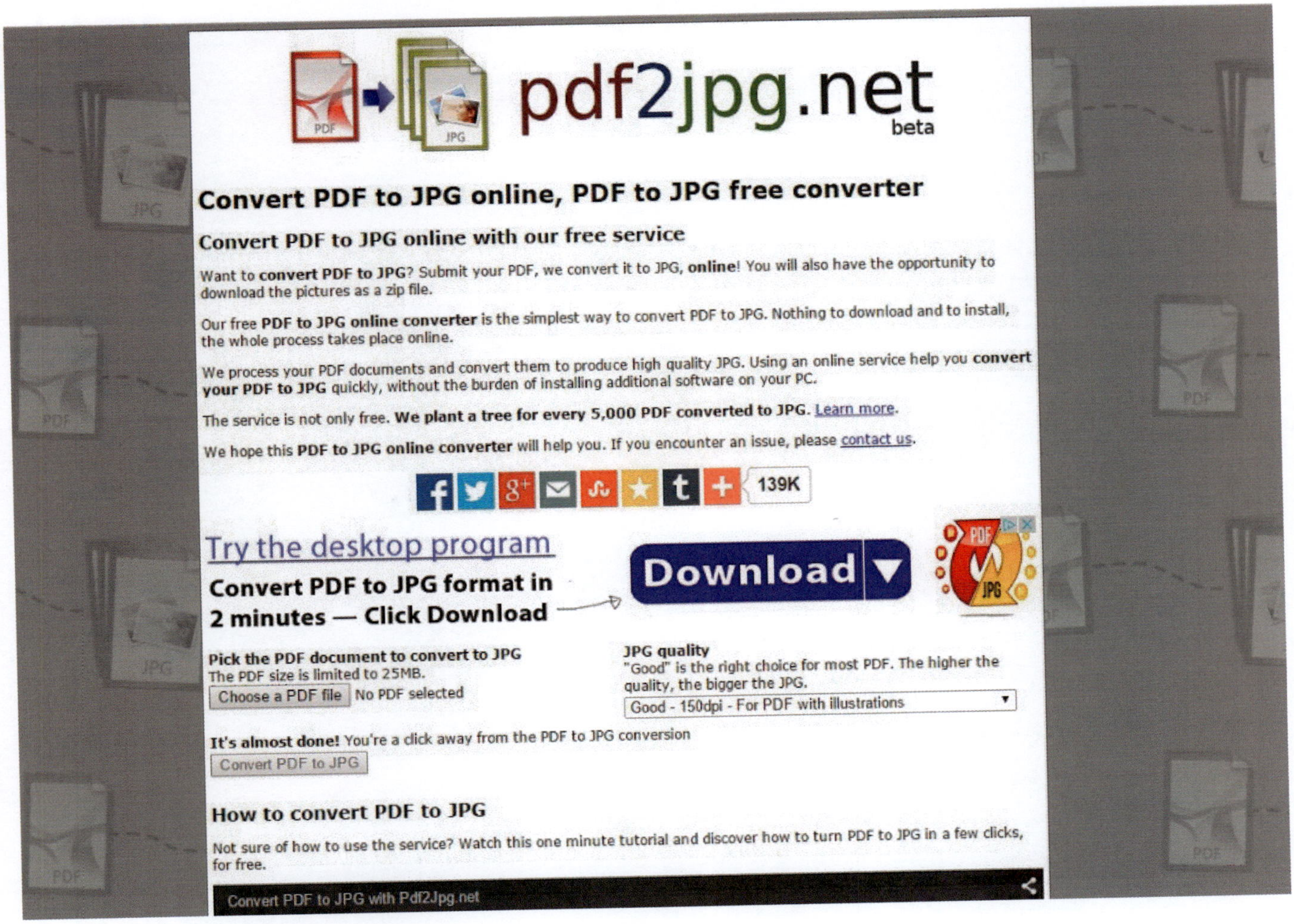

Convert PDF to JPG online with pdf2jpg.net free service.
You can also have the opportunity to download the pictures as a zip file.

This free **PDF to JPG online converter** is the simplest way to convert PDF to JPG. Nothing to download and to install, the whole process takes place online.

Using an online service help you **convert your PDF to JPG** quickly, without the burden of installing additional software on your PC.

44 - WeTransfer.com

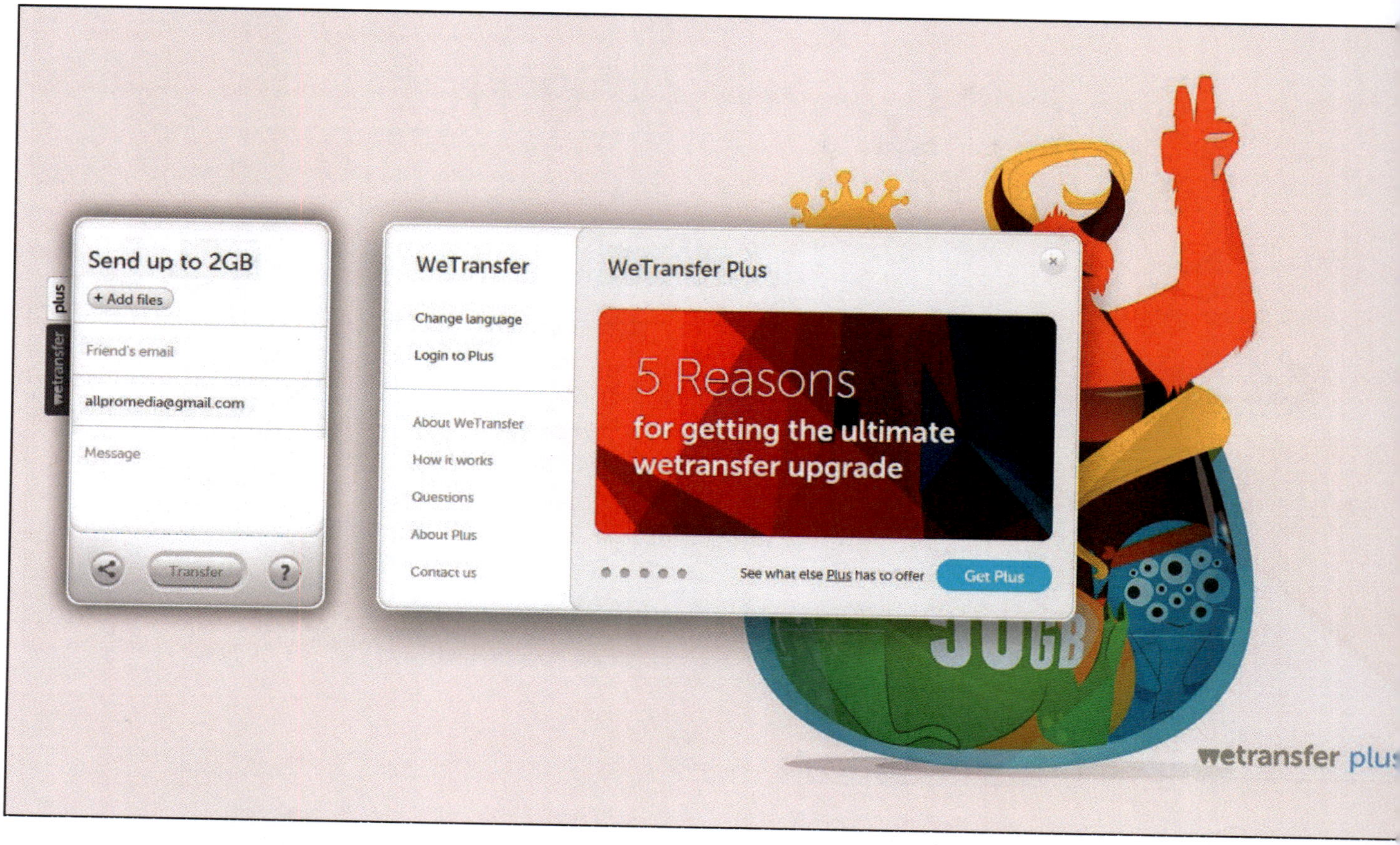

WeTransfer is an online file-transferring platform, which basically means that you send it, **we transfer**. There's no hassle. No stress. No charge.

WeTransfer is a cloud-based file transfer service based in Amsterdam designed to send small to large files. WeTransfer users can send files up to 2GB in the free version.

45 - HundredZeros.com

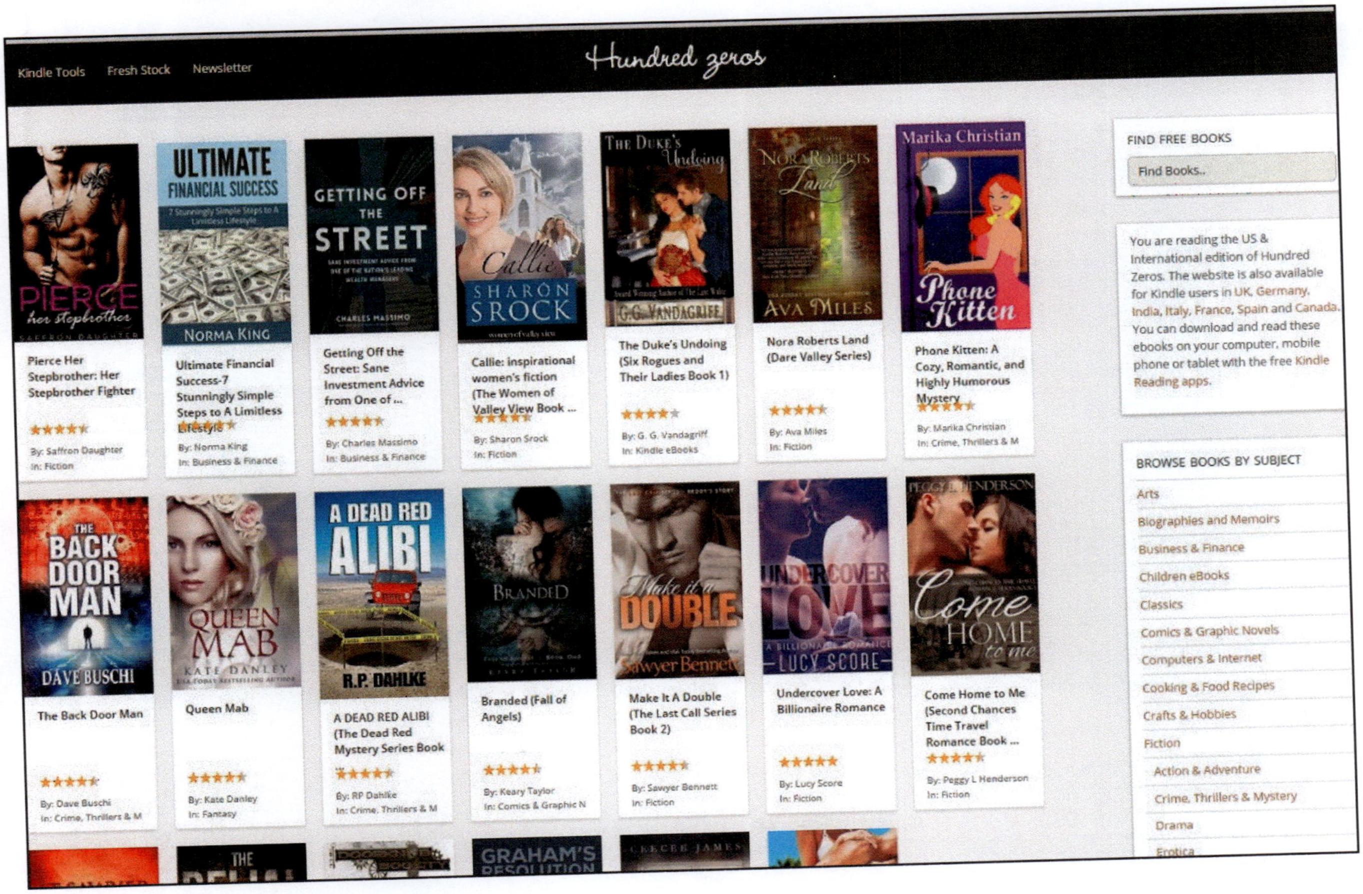

Hundred Zeros catalogs piles of free Kindle books. You can browse the front cover page for the current top books, browse by category, or search by topic in the sidebar. When you find a book you like just click through to Amazon and send to your Kindle or Cloud Reader.

46 - Pixlr.com

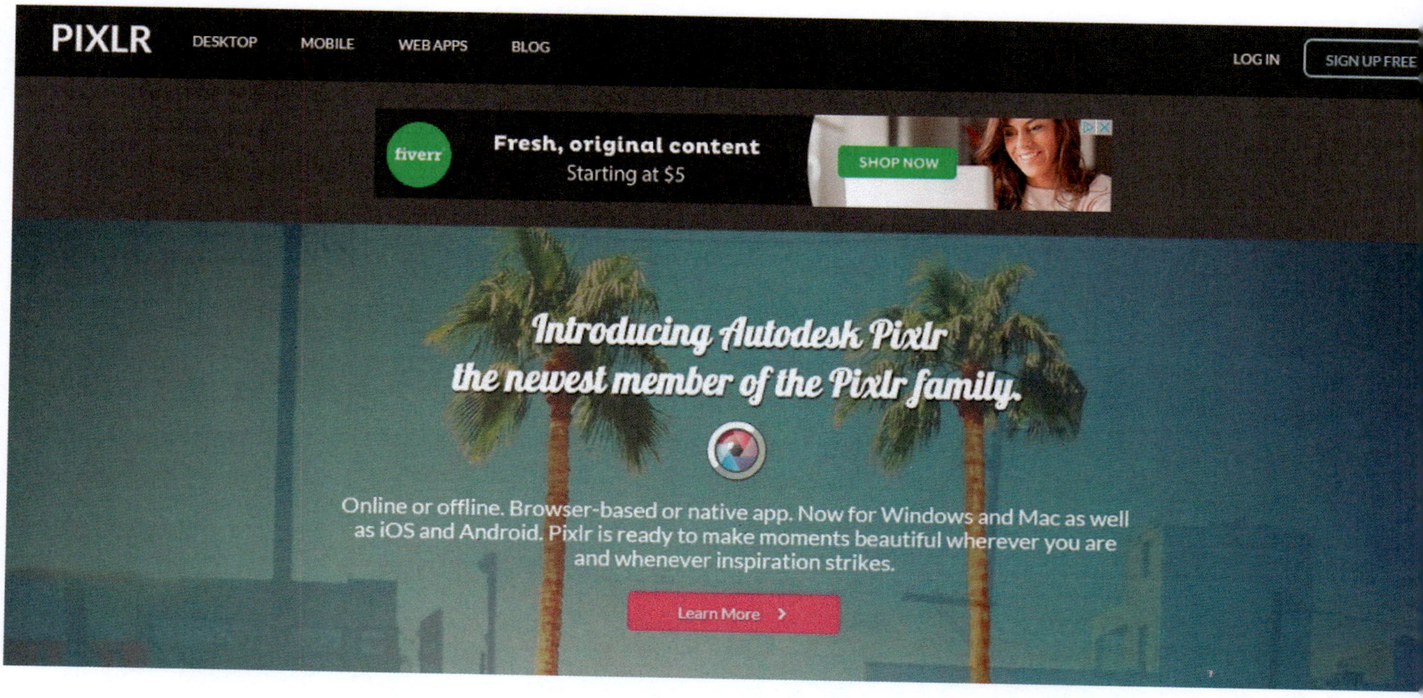

Pixlr is a cloud-based set of image tools and utilities. It includes a number of photo editors, a screen grabber browser extension, and a photo sharing service. This service was intended for non-professionals, however the apps range from simple to advanced photo editing.

Think Adobe Photoshop but for free and in the cloud!

47 - Lynda.com

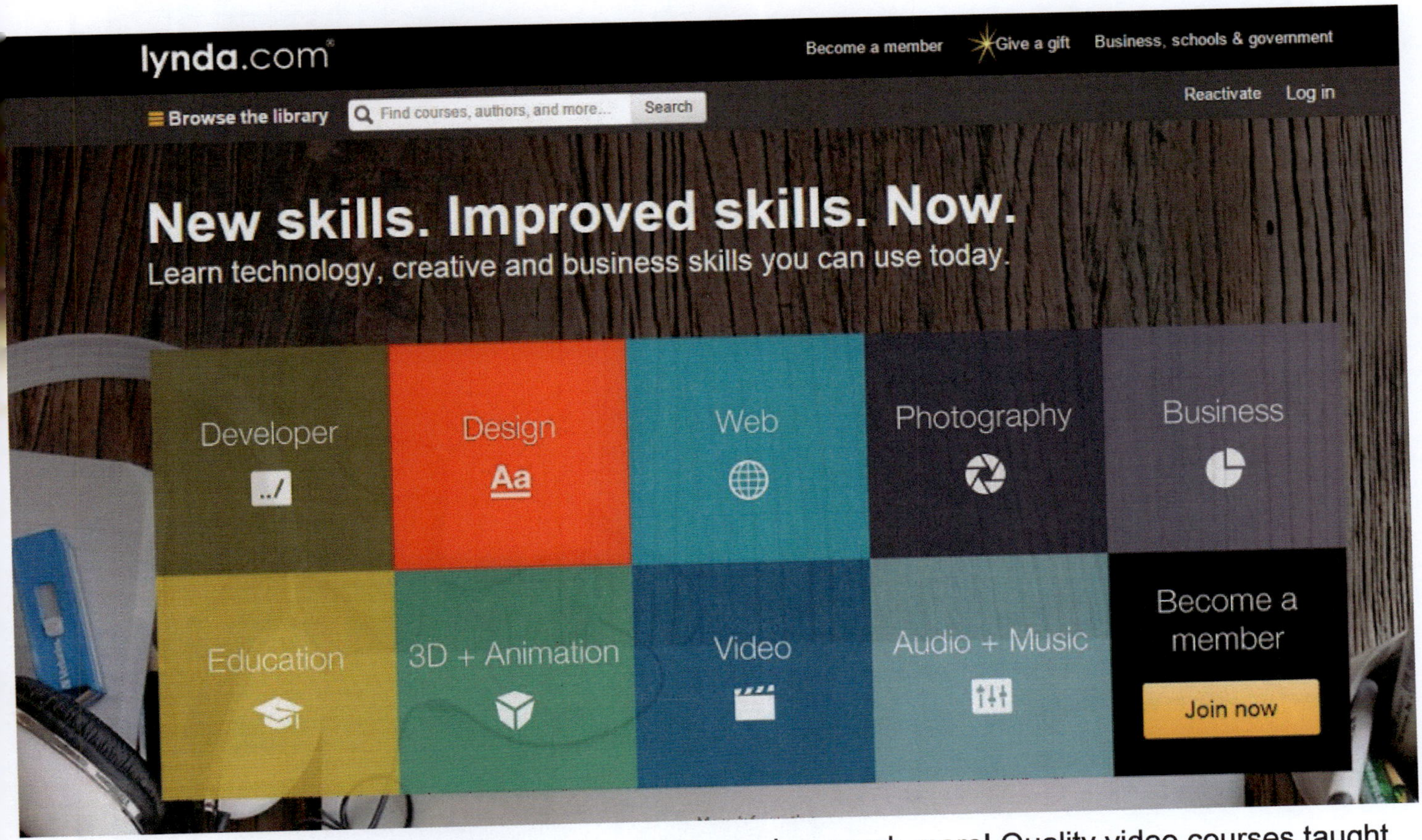

Learn about software from Microsoft, Adobe, Apple, and so much more! Quality video courses taught by industry experts

Lynda.com is a leading online learning company that helps anyone learn business, software, technology and creative skills to achieve personal and professional goals. Members have access to the **lynda.com** video library of engaging, top-quality courses taught by recognized industry experts. **Lynda.com** also employs expert instructors that are true masters in their fields and who are also great teachers as well. They can answer all your questions and will offer useful advice. Their mission is to impart knowledge regarding correct workflow and they will also teach you how to develop skills required to solve problems. The best part is that none of the teachers at **Lynda.com** will ever talk down to their students.

You can choose between single-user subscription plans and multi-user subscription plans. You can also try their services for free before becoming a subscriber.

Lynda.com is dedicated to helping people learn, master and also apply various digital tools as well as techniques in the best manner possible.

savvy INVESTORS .com

48 - DealCatcher.com

Catch the best deals, coupon codes and printable coupons for Amazon, Kohls, Target, Best Buy, Home Depot and thousands more, updated daily.

DealCatcher is a very appealing hot deals website with great bargains to be found by all discount shoppers. Deals can be had on many items listed on the site including computer systems, clothes, jewelry, and other types of items. Visually, **DealCatcher**.com is very well designed with easy to navigate buttons and an easy to read format. The main page displays a general list of current deals on the site. Sections on the site include printable coupons, Sunday ads, price comparison, and a forum. A Black Friday section is also available for end of November deals. Many coupon codes are integrated into the list of general deals so shoppers do not have to seek out each code separately from the deals section. Shoppers can browse deals based on the categories of each type of item.

49 - GrassHopper.com

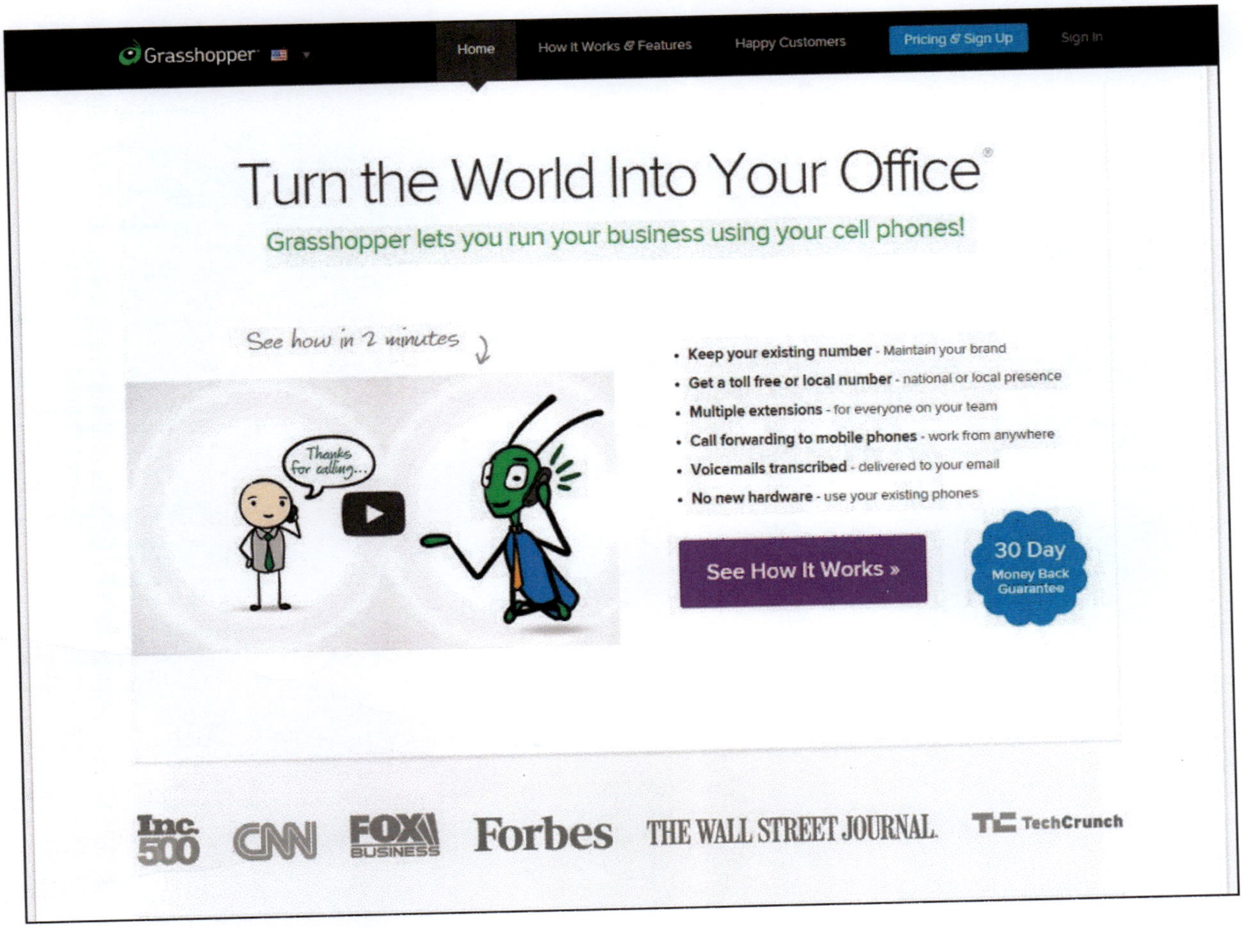

Grasshopper (Phone System) is a business that sells call forwarding and answering services to small businesses.

Grasshopper lets you run your business using your cell phones! Keep your existing number and maintain your brand; Receive a toll free or local number.

Grasshopper is a virtual PBX service that ties together existing phone lines by using the power of the web to connect mobile phones, land lines and VoIP accounts. It offers local as well as toll free numbers for long distance incoming calls, call forwarding to landlines and mobile phones. There are 800 numbers and unlimited extensions at a highly reasonable rate. Arguably, the best thing about **Grasshopper** is that it requires no additional hardware to get started. You can add existing phones to your plan at will, manage your account online and extend your voice phone system as far as you need to.

50 - RelayRides.com

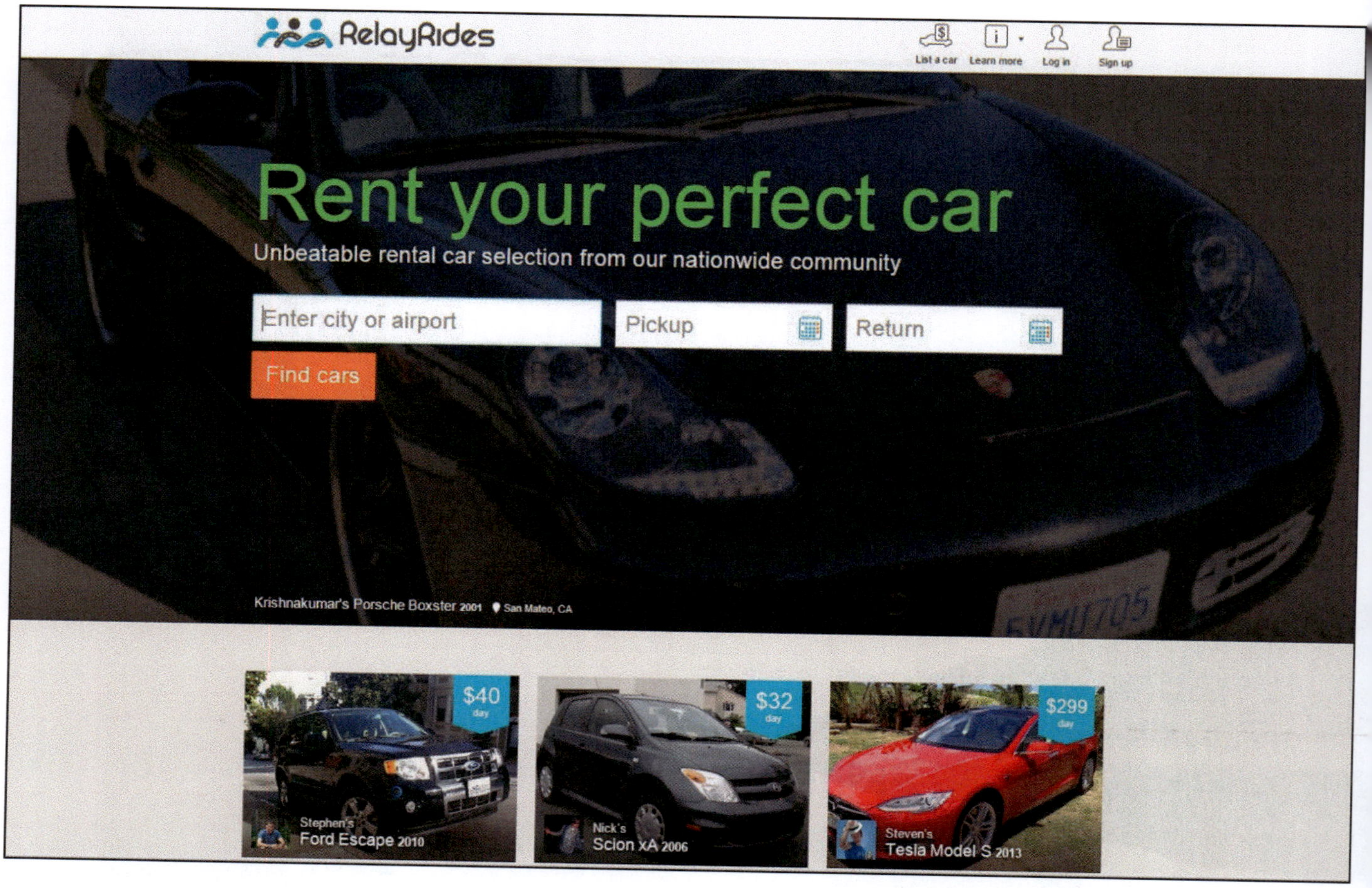

RelayRides is a Nationwide peer-to-peer carsharing marketplace. You can rent a car, or make money by renting out your vehicle through their online interface. Car owners can set their own prices and the company takes 25%.

If you're involved in a collision in your vehicle, rest easy — you're covered by a $1 million liability insurance policy that protects you against claims from third parties for injuries and property damage. You're also protected against physical damage to your car (up to the actual cash value) for losses due to collisions. Most "comprehensive" causes (fire, theft, etc.), with some specific limitations for acts of nature, are protected.

51 - Bill.com

Bill.com's mission is to dramatically simplify how small and mid-sized businesses pay their bills and collect payments. **Bill.com** turns the endless hours and administrative drudgery of traditional paper-based payment processing into a one-click experience. **Bill.com** is using cloud technologies and its industry-leading expertise to radically streamline business payment processes.

52 - Google Calendar

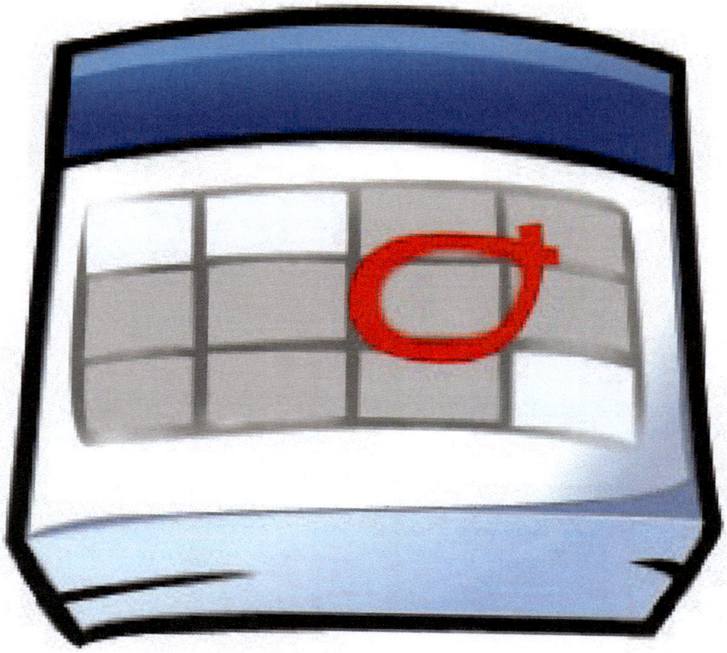

Google calendar

With **Google's** free online calendar, it's easy to keep track of life's important events all in one place. Scheduling is easy if you know when everyone is free or busy. Share your schedule to let your co-workers, family, and friends see your calendar, and view schedules that others have shared with you.

Google Calendar is a time-management web application and mobile app for created by Google. It became available on April 13, 2006, and exited the beta stage in July 2009. Users are required to have a Google Account in order to use the app.

I love the feature that you can set it up to text(sms) you to remind you of your up and coming scheduled entry!

53 - LendingClub.com

Lending Club is the world's largest online marketplace and is peer-to-peer lending company connecting borrowers and investors. **Lending Club** is transforming the banking system to make credit more affordable and investing more rewarding. **Lending club** operates at a lower cost than traditional bank lending programs and pass the savings on to borrowers in the form of lower rates and to investors in the form of solid returns.

Why Real Estate Investors? It's like rental property. If you have cash you rent it out and you make a return, as the website says, about 8% after all the fees and defaults.

54 - Prosper.com

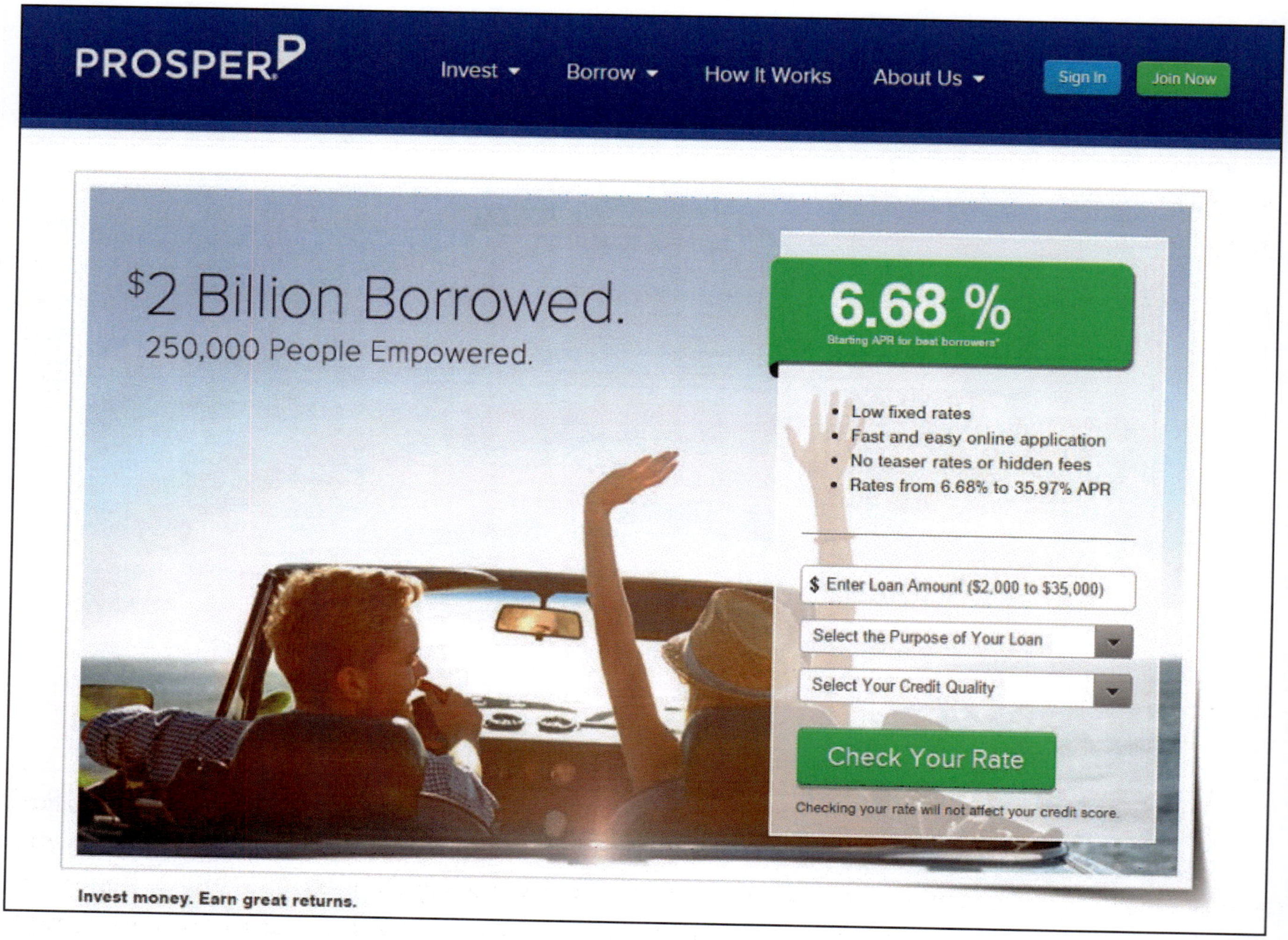

Prosper is a peer to peer lending company that offers personal loans at low rates. These loans are unsecured, which means you do not have to put up any collateral (like a house or car) that could get taken away if you can't make payments. Each loan is typically funded by multiple people all over the United States.

Individuals can either invest in personal loans or request to borrow money.

55 - eVoice.com

eVoice.com manages incoming and outgoing calls using a virtual phone number.

When customers call, they are connected through eVoice. Callers hear customized and professionally recorded greetings; and are then routed where you want, when you want, based on rules you create. You'll never miss an important opportunity and can avoid unwanted or unimportant calls.

56 - Elance.com

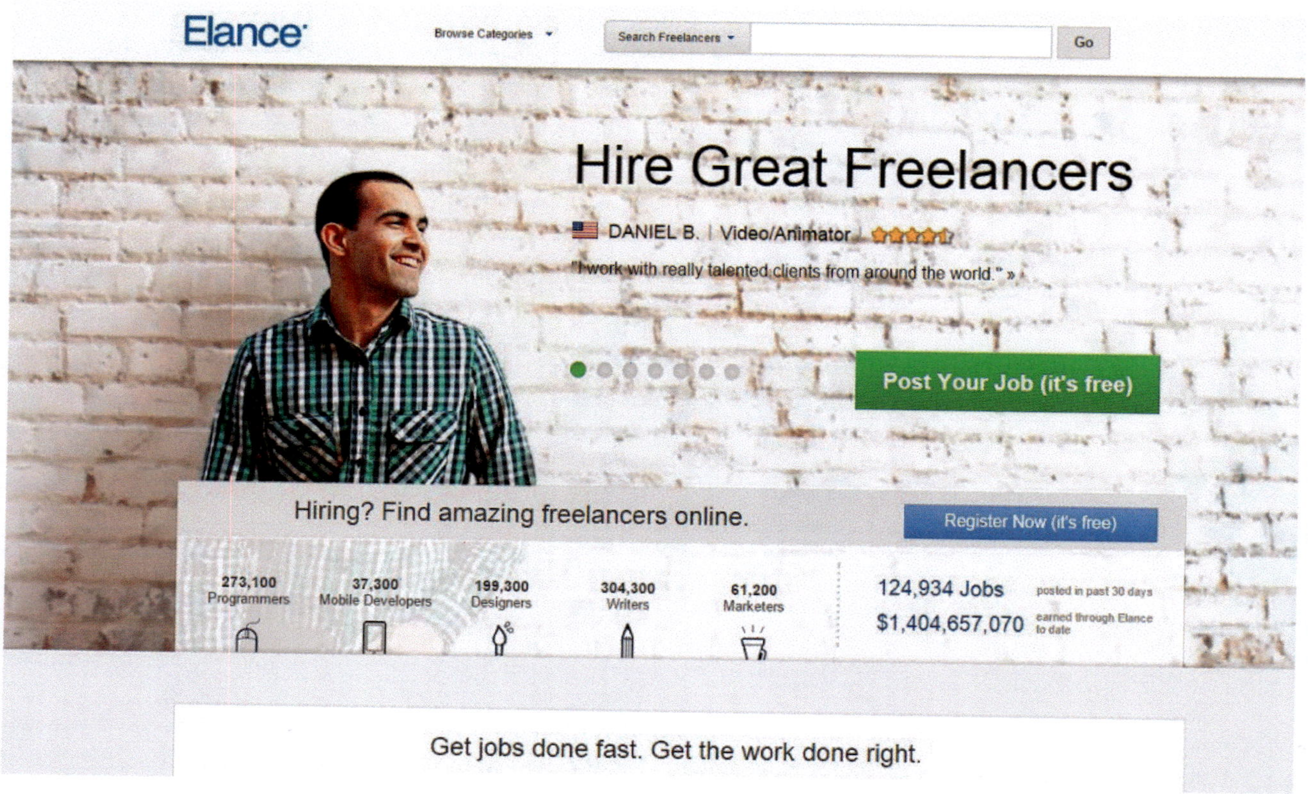

Elance is one of a virtual site where freelancers and project owners meet to get the job done.

For every job, you pay only for work you've received and then approved. If there's a disagreement between you and your freelancer, we'll provide free dispute assistance and arbitration. You can use your MasterCard, Visa, American Express, PayPal or an **Elance** Account to pay freelancers.

57 - oDesk.com

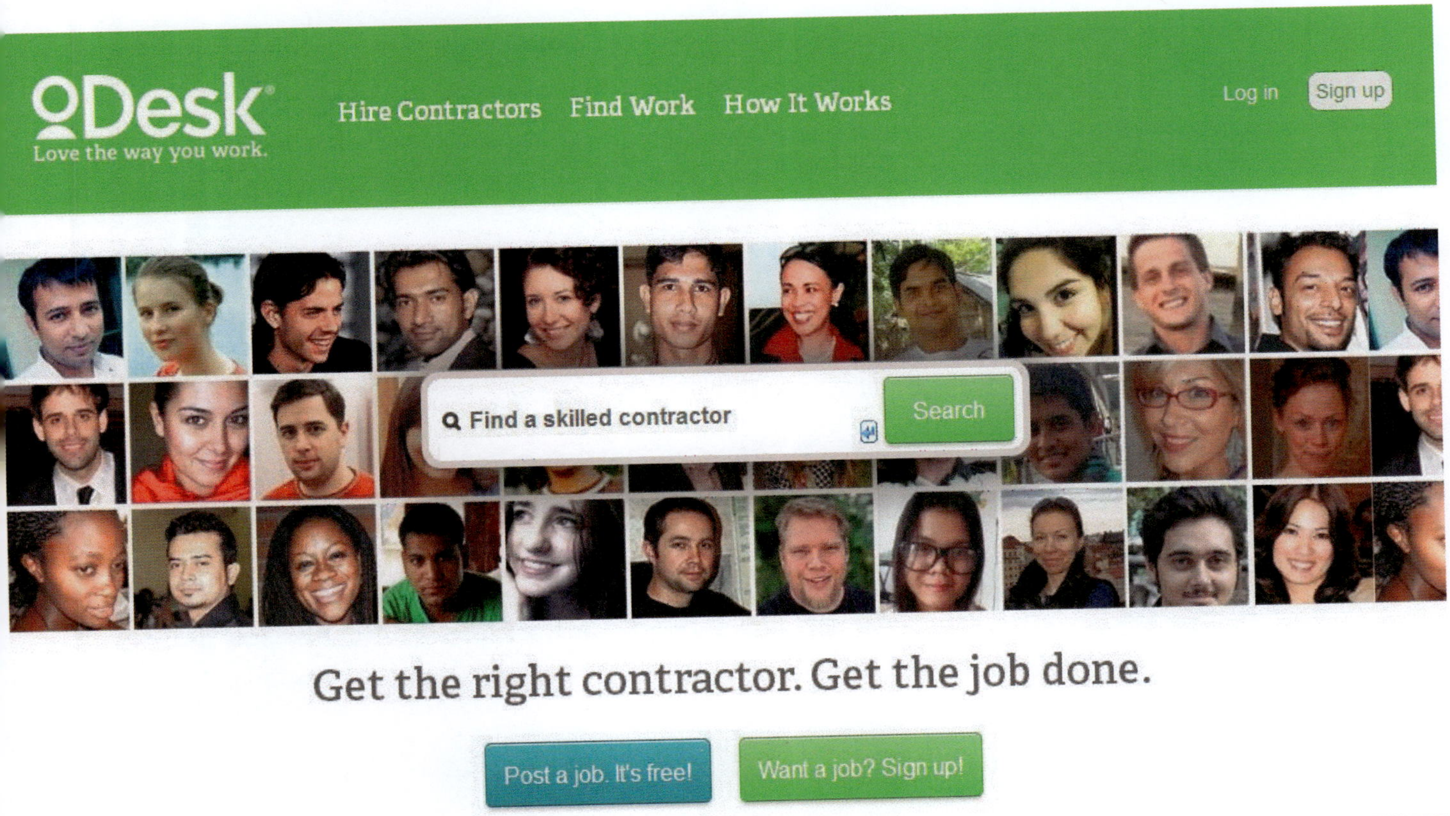

Anything you can do on a computer—from graphic design to software development—can be done on **oDesk**. **oDesk** freelancers can tackle a wide range of projects.

Find freelancers and freelance jobs on **oDesk** - the world's largest online workplace where savvy businesses and professional freelancers go to work!

In a nutshell, **oDesk** is an innovative platform that connects employers with virtual freelancers. Aside from the cost-saving benefits of hiring hourly employees, businesses also appreciate the opportunity **oDesk** presents to reach talent from around the world.

savvy INVESTORS .com

58 - Mashable.com

Mashable is a leading source for news, information & resources for the Connected Generation.

Mashable reports on the importance of digital innovation and how it empowers and inspires people around the world.

Mashable has become synonymous with social media. The website is a constant presence in top rankings and has become the most dominant social media news outlet on the planet. Tech enthusiasts, digital media nerds, and everyone from social web newbies to seasoned "experts" read it and respect its position. It's difficult not to. In 5 years time the blog has become nothing less than a powerhouse of information and influence.

59 - Hive.im

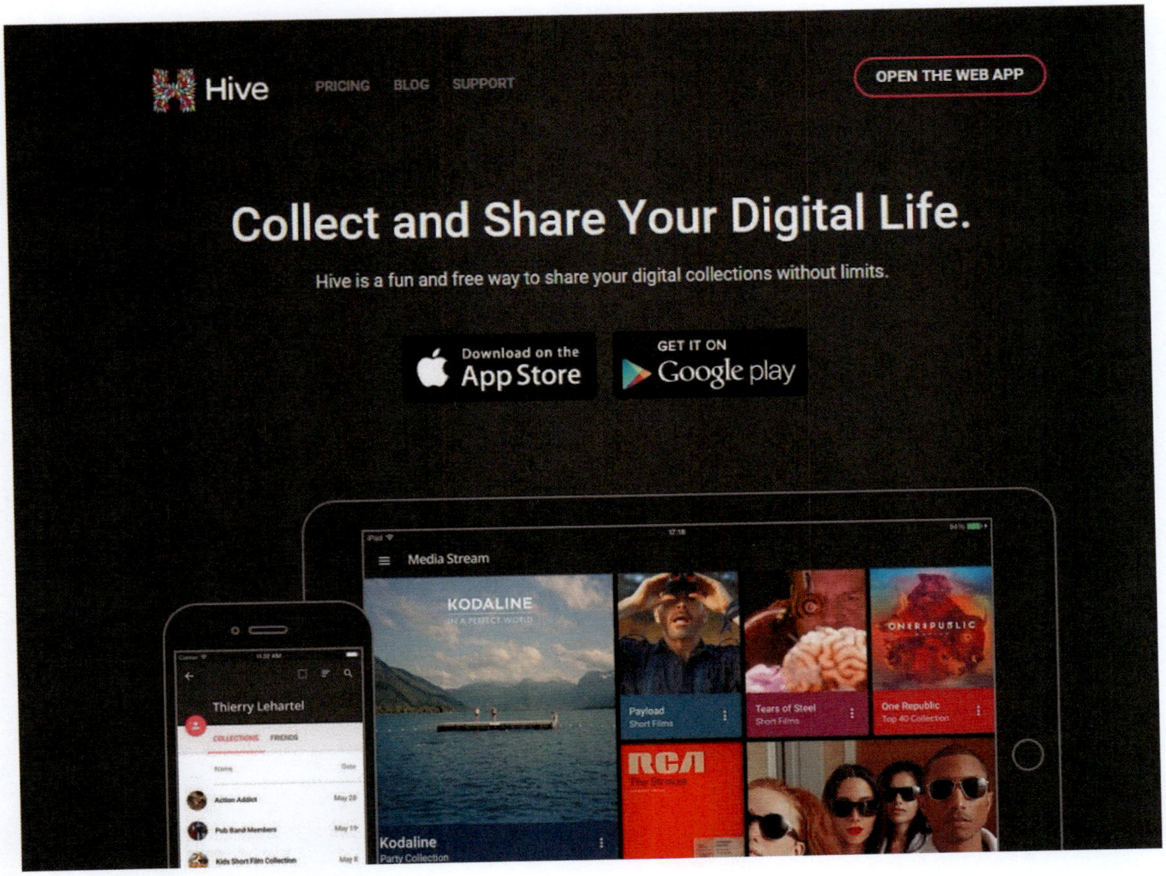

Hive is a free, unlimited cloud storage space to share your media, photos and files with friends.

For the privacy conscious: **Hive** is ad-supported, and reserves the right to screen your files, so look elsewhere for storing private files.

These things make **Hive** a good place to dump neat, innocent things for your friends and family to access, but not a good place for confidential, intimate, or work materials.

Hive is ideal for social file-sharing between friends. **Hive** could be a great place to put videos of your band's jam sessions for friends to stream, and also photos for relatives to download.

Is there a place for **Hive** in your life? A spot in between "I'm going to put this up on Facebook" where files quickly fall off your friends' newsfeeds, and "I'm going to put this file in Dropbox/Google Drive" where files can only be accessed and edited by select individuals?

60 - BackPage.com

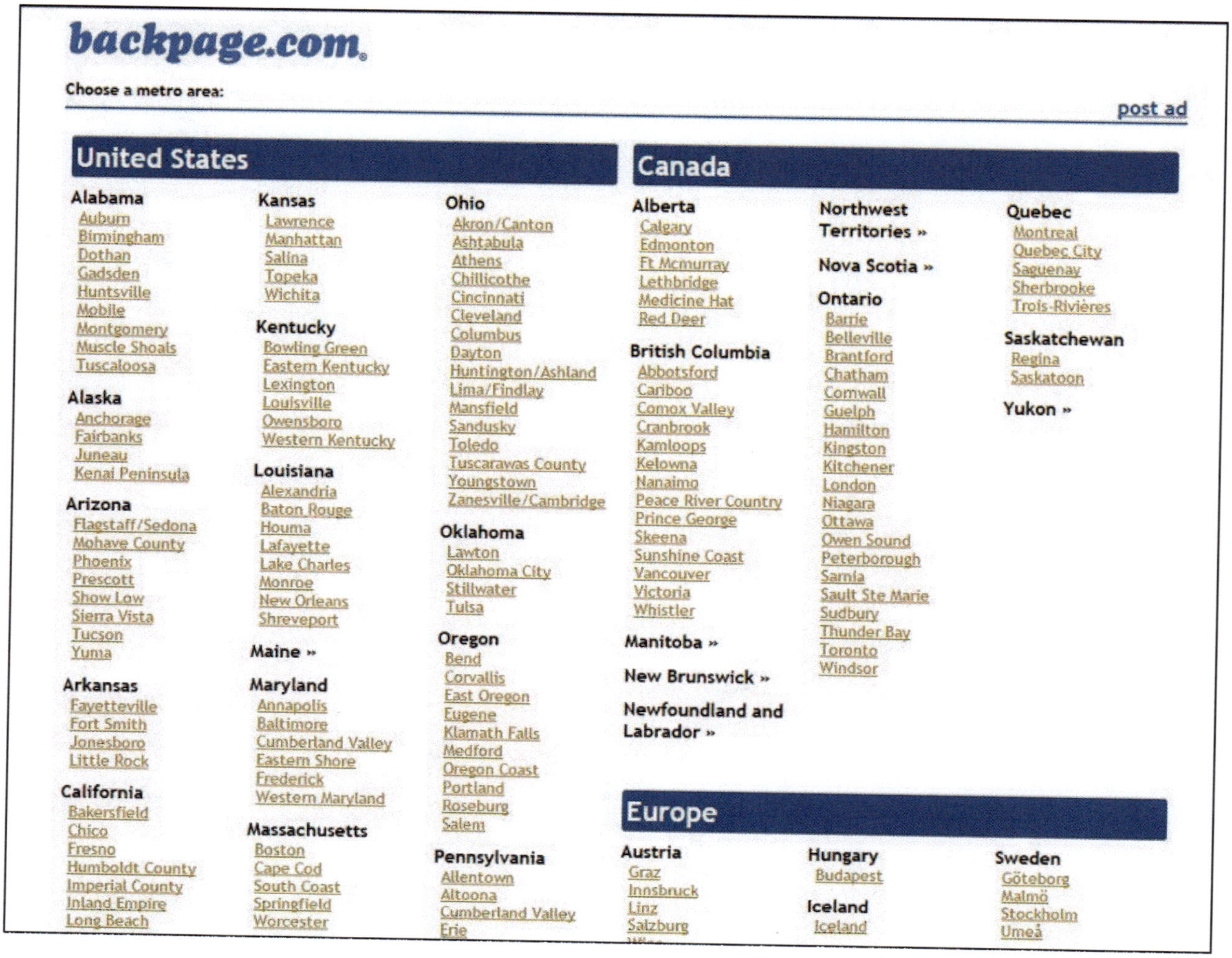

Backpage is a classified advertising website. It offers a wide variety of classified listings including automotive, jobs listings, and real estate. **Backpage** is the second largest classified ad listing service on the Internet in the United States after Craigslist.

savvy INVESTORS .com

61 - SkillShare.com

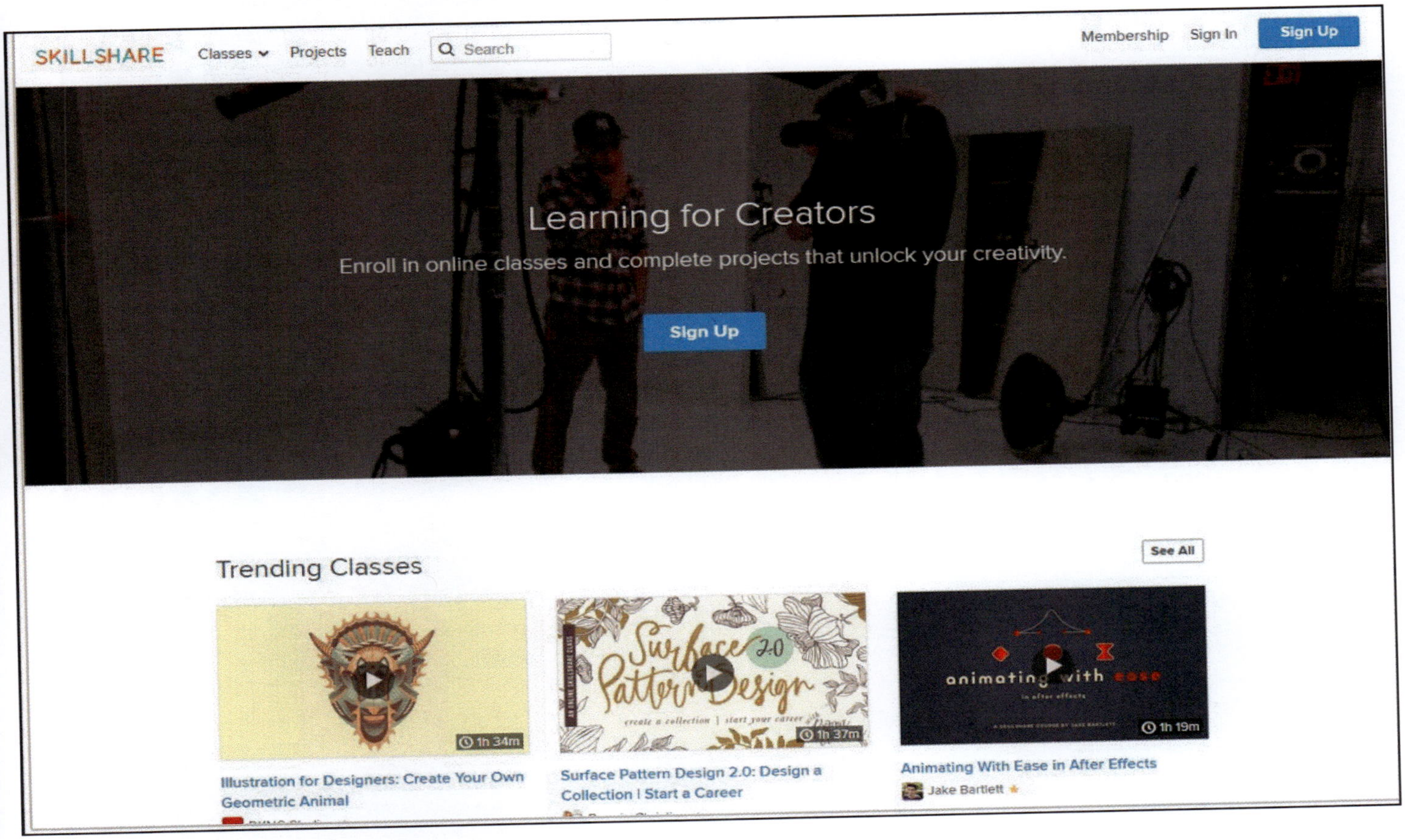

Skillshare is a learning community for creators. Anyone can take online classes, watch video lessons, create projects, and even teach a class themselves. Their on a mission to unlock creativity around the world.

You'll have access to online classes that you can watch on your own time, and you'll get to collaborate with **Skillshare** vibrant and active community on student projects.

62 - Rev.com

Rev.com is a new breed of service provider. Like traditional vendors, **Rev** rely on a network of hand-picked freelancers. Unlike traditional vendors, **Rev** uses technology to offer unbeatable quality, speed, and value.

Rev's proprietary technology platform is built to create a great work environment for transcriptionists, captioners, and translators to deliver clients ever increasing speed and quality.

Rev's mission is to give more people the freedom to work from home. **Rev** is bringing the best of the office to an online workplace. They believe that attracting the best workers is the key to delivering great service to our customers.

63 - Life360.com

Life360 is a family networking app launched in 2008. It is a location-based service that allows family members to share their location with each other, and easily communicate throughout the day. The app has more than 50 million registered families and has received numerous awards.

The app also populates the location of registered sex offenders near your location or your loved ones location.

64 - SlimVoice.co

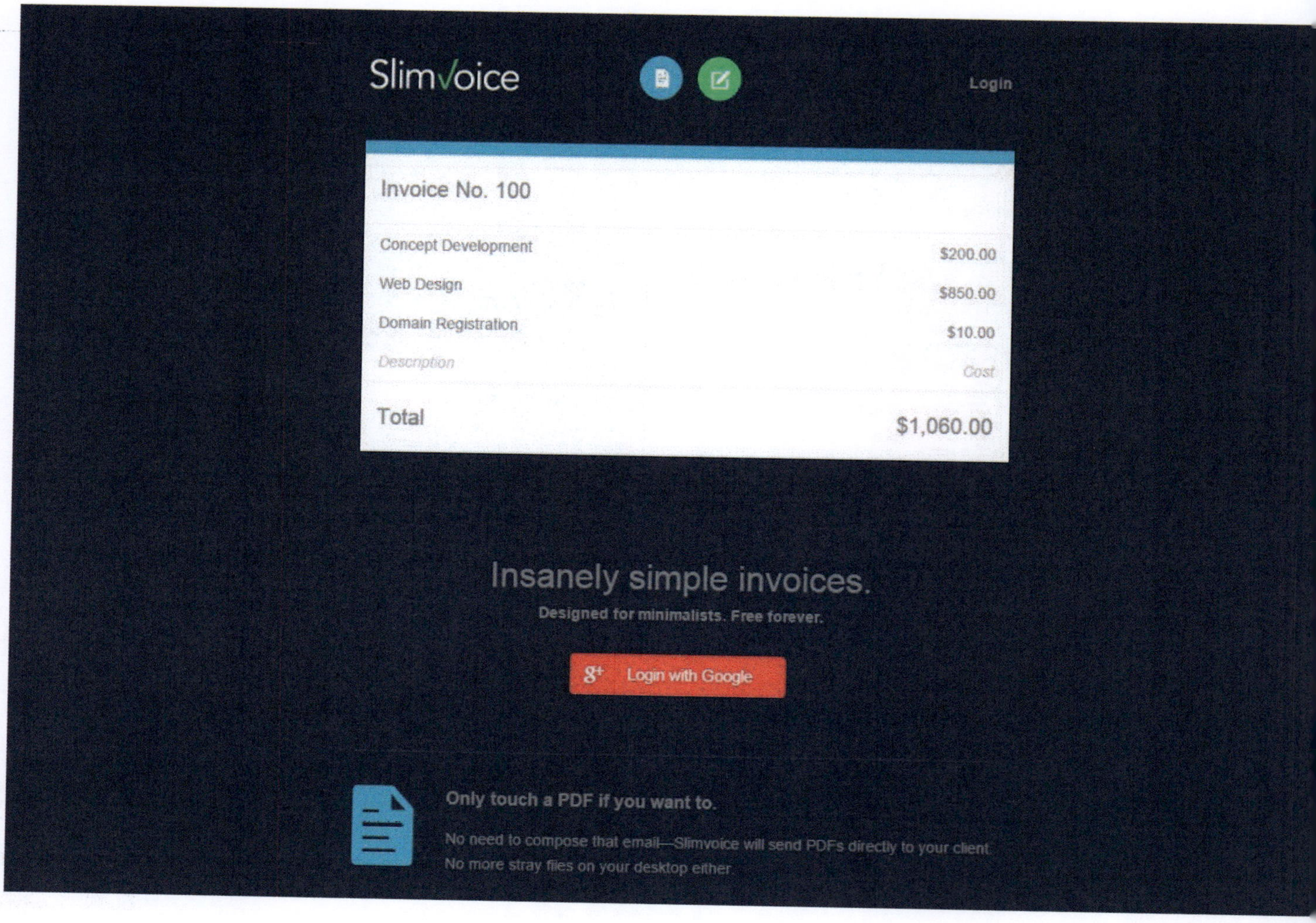

Create invoices for free. You can save unlimited invoices for free by creating an account.

65 - TrueCar.com

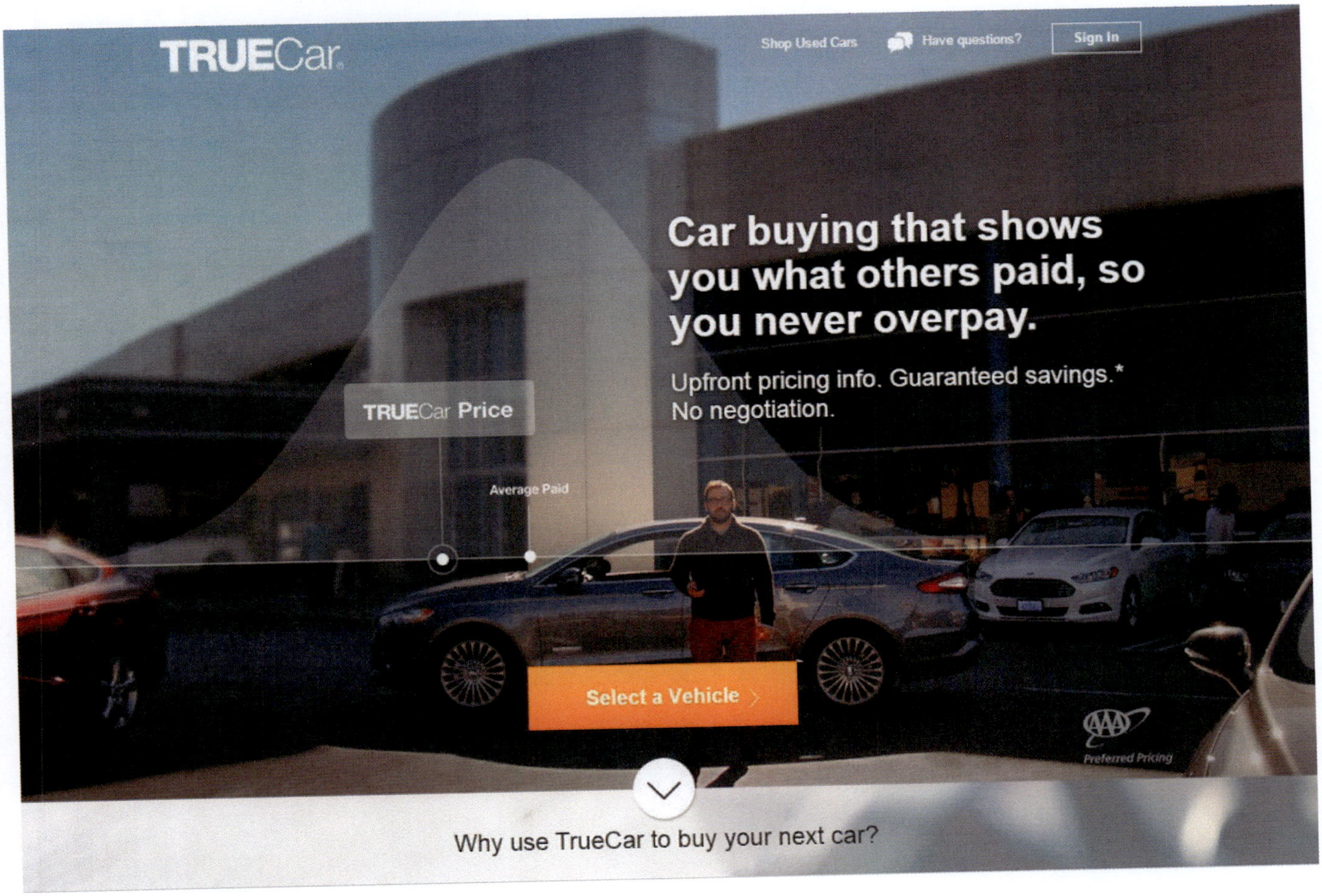

TrueCar is an information and technology platform that enables its users to communicate with dealers for a hassle-free car-buying experience.

TrueCar is an automotive pricing and information website for new and used car for buyers and dealers. The company gives buyers information on what others have paid for cars, upfront pricing, and a network of **TrueCar** dealers.

TrueCar is a great concept: you can figure out a price for your car and trade-in online, without any of the frustrating negotiations.

savvy INVESTORS .com

66 - Amazon Prime

Amazon **Prime** is a paid service that gives Amazon shoppers a few distinct advantages. Members of Amazon **Prime** are eligible for free two-day shipping on any item, without a minimum-order balance. Otherwise, free shipping requires a $35 minimum purchase. The service also gives members access to one free Kindle book to borrow (from a pool of about 350,000 selections) each month, lets them use the streaming video service "Amazon **Prime** Instant Video" and allows them to stream music via Amazon **Prime** Music.

Cost is $99 a year.

67 - Dymo LabelWriter 450

The **LabelWriter** 450 Twin Turbo label printer holds two label rolls, so you can print labels and postage even faster. Label, mail, and file smarter with proprietary DYMO® Productivity Software, featuring the new DYMO Label™ v.8, DYMO Stamps®, DYMO File™ and Address Fixer™.

DYMO Label v.8 software enables you to create and print address, shipping, file, folder and barcode labels, plus name badges and more. With **LabelWriter** you can do it all without the extra work and hassles of printing sheet labels on a standard desktop printer.

68 - PaperPort

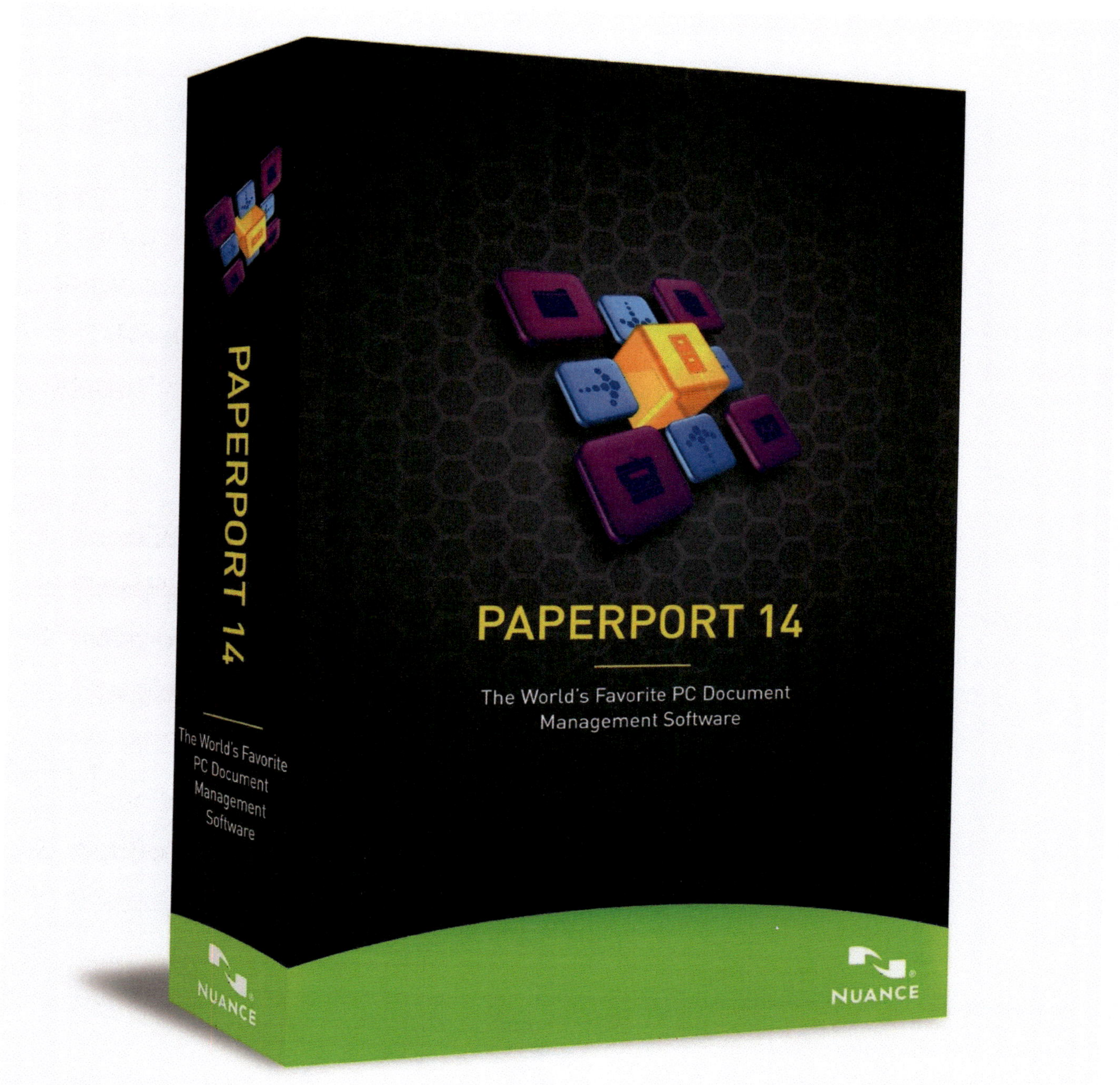

PaperPort 14 is the fastest and easiest way to scan, share, search, and organize documents, files, and photos. **PaperPort** provides a single way to scan paper, view, edit, and convert files to access on your PC.

savvy INVESTORS .com

69 - Fujitsu ScanSnap S1300i

At nearly half the size of a sheet of paper, the PC and Mac compatible **ScanSnap** S1300i is the smallest multi-page, double-sided scanner for one-button ease of use at the desk or on the road. Stack up to 10 pages into the automatic document feeder (ADF) and press the blue button.

ScanSnap S1300i scans both sides of a page at up to 12 pages per minute directly to PDF, searchable PDF, JPEG, or to office applications like Word or Excel®. The **ScanSnap** S1300i automatically recognizes the size of each document, detects and corrects for skew, and shows images in their proper orientation. You can even choose to have blank pages automatically removed.

savvy INVESTORS .com

Making a Difference in Kids' Lives

After School Tutoring and Mentoring Program

The mission of Whiz Kids is to improve the well-being of inner-city youth through academic tutoring and positive mentoring relationships.

About Whiz Kids

Founded in 1996, under the umbrella of the 501(c) 3 nonprofit, City Care, Inc., Whiz Kids is a nonprofit, one-on-one, tutoring and mentoring program for at-risk students in the Oklahoma City metropolitan area. Whiz Kids collaboratively and creatively addresses Oklahoma City's urban social and educational crisis by matching tutor/mentors with at-risk students. Teachers select students for the Whiz Kids program because they are reading below their grade level and can benefit from a mentoring relationship. Tutors are paired with a child and meet weekly with the same child during the school year for a one-on-one after-school tutoring session. Whiz Kids sites are churches in each school's community that donate their facility and provide tutors and on-site coordinators. These urban churches are partnered with other metro area (typically suburban) churches to provide additional volunteers and tutors, essentially bringing the resource of people back to the urban neighborhoods where the children live. Each site has a liaison teacher from the school that serves as an educational consultant, and is a resource for the volunteer tutors.

In order to qualify for the Whiz Kids program, a school must have at least 87% of its students eligible for free/reduced lunch. Seventeen of the Whiz Kids schools have percentages higher than 95% free/reduced lunch and four are 100% free/reduced lunch. Because of this, Whiz Kids serves children who live in some of Oklahoma City's most desperate neighborhoods, where poverty, illiteracy and violence often strangle a child's promise. The schools served by the Whiz Kids program are mostly Oklahoma City Public Schools, but also include a few schools in the Crooked Oak, Putnam City, Mid-Del and Crutcho districts.

Since its inception, one school and a handful of volunteers, Whiz Kids has grown to serve more than 800 students from 29 at-risk elementary schools, in partnership with 63 churches.

Get involved and support this needed impactful program!

Whiz Kids
OKLAHOMA
A Division of City Care

www.whizkidsok.org

Hello **Savvy Investor!**

How do you know what material is right for you? Great question! Below you will find my recommended materials for any level of investing. As the Savvy Landlord, I understand what it was like to be a beginner with no resources available to me. We have something for everyone.

Gain Motivation as an Investor

- Provides Clear Insight on Deals
- Examples of Investment Failures and Successes
- Relatable Stories of What Investing is All About
- Real Deals with Addresses and Authentic Mistakes

"As a real estate investor and landlord, I have read dozens of books on real estate investing. Reading The Savvy Landlord has not only been an eye opening experience, the book also serves as a reference for me." ~ Marshal Terry (Amazon)

Answer the Questions all Investors have

- Questions About Tenants to Evictions
- Relevant to Any Investor
- Discover the Answers to Those Hard Questions
- Comprehensive Handbook

"I thought this book was great. All content no fluff, and real life information."
~ Greg Meech (Amazon)

Conquer the task of Hiring Contractors

- The Art of Handling Contractors
- Scams & Pitfalls to Avoid
- Proven Systems to Save Time & Money
- Terminology Defined

"Great intuitive book for anyone that is interested in real estate investing."
~ Anonymous (Amazon)

Graphic Novel: Dare to Dream

- Inspires & Gives Hope
- Empower Readers of All Ages
- Rags To Riches Story
- Build Self-Esteem

"Great for kids and adults alike. Tells the amazing story of one man's journey towards attaining his dream.." *~ Jessica Watson* (Amazon)

Secret to my success is reading! Try to take some time, as little as 10 minutes a day to read so you will learn and grow. *Remember*, the more you know the more you make.

SavvyInvestors.com | 405.371.2299

Printed in Dunstable, United Kingdom

91335486R00045